TSUNAM

THIRTY-TWO

TSUNAMI!

Second Edition

Walter C. Dudley
Min Lee

A Latitude 20 Book
University of Hawai'i Press
Honolulu

Printed in the United States of America

03 02 01 00 99 98 5 4 3 2 1

Library of Congress Cataloging-in-Publication Data

Dudley, Walter C., 1945–
Tsunami! / Walter C. Dudley, Min Lee. — 2nd ed.
p. cm.
Includes bibliographical references and index.
ISBN 0–8248–1969–1 (alk. paper)
1. Tsunamis—Hawaii. I. Lee, Min, 1943– . II. Title.
GC222.H3D84 1998
363.34'9—dc21 98–24953
CIP

University of Hawai'i Press books are printed on acid-free paper and meet the guidelines for permanence and durability of the Council on Library Resources.

Designed by Nighthawk Design

Contents

Foreword

The ocean is always with us in the Hawaiian Islands. It surrounds us, brings us rain, gives us food, and provides a never-ending source of pleasure and beauty. But its power must be recognized. Those of us who live in Hawai'i know the importance of the ocean to our way of life. We must also acknowledge its might.

Tsunamis are a manifestation of the ocean's power that should never be forgotten. In 1946 and 1960, people in the islands, and those in Hilo in particular, were reminded of the giant force latent in the sea. Many will remember with me the shock at the devastation, the sorrow at the losses.

Accounts of these tsunamis were published soon after the events, but it was not until 1986, when Walter Dudley and Min Lee wrote the first edition of this book, that the subject of Hawaiian tsunamis was treated in a comprehensive way, revealing the background facts and expressing the human reality.

In the intervening years, more survivors have come forward with their stories, more knowledge has been gathered, and many technical advances have been made. This new edition gives even more background about the Hawaiian tsunamis, but it also expands the horizon to include experiences throughout the Pacific and beyond.

The authors have gathered together many amazing stories of tsunamis, and they make fascinating reading. But this book is more than a collection of tales. Walt Dudley uses his specialist knowledge to set out clearly and accurately the why and how of tsunamis. He explains the development of the Tsunami Warning System and the way in which it works today.

The events of the evacuation during the Hawai'i tsunami of May 7, 1986, underlined the fact that everyone needs to be informed about tsunamis and aware of their destructive capability. The practical usefulness of this book is one of the ways it appeals to me. Another aspect is the good reading it provides.

We read about people, in Hawai'i and across the Pacific, in times of great stress. Their stories illustrate human resilience, the ability to accept what fate has to offer and to start again. In 1946 and 1960, my company, Hawai'i Planing Mill, Ltd., was leveled by tsunami waves. Like other businessmen in Hilo we rebuilt, each time a little farther from the ocean!

In this new edition the authors show us on an even larger canvas how natural forces act and how people react; we can learn and enjoy at the same time.

Robert M. Fujimoto
Big Island business owner who has known the power of the waves

Preface

Tsunamis remain a true and constant threat to the people of Hawai‘i. The Tsunami Warning System has done much to alleviate this danger, and, if the warnings are heeded and understood, people need not live in fear of the waves. It was with this idea in mind that we wrote the first edition of this book. It is our hope now, as it was then, that the personal experiences shared herein, together with the explanations of how these giant waves behave, will help those in Hawai‘i and throughout the world to be better prepared for the next tsunami.

The first edition focused primarily on the tsunami threat to the Hawaiian Islands, but in this edition we have expanded the scope of the book to include tsunamis around the world. We have also included a simplified summary of recent tsunami research findings. This would not have been possible without the dedicated work of the scientists who study tsunamis and to whom we are indebted. The reference sections at the end of this book list the research and historical articles we have relied on for much of the information herein. We have made every effort to be as accurate and up-to-date as possible and any errors or omissions are unintentional.

Acknowledgments

I would like to give special thanks to the many people who have given generously of their time to assist me with this work. I thank Dr. Doak Cox, "the father of tsunami research," for his encouragement. I am enormously grateful to Dr. James Lander and Ms. Patricia Lockridge, the real pioneers in collecting information on tsunamis in the United States, whose exhaustively researched books on tsunamis made my job easy and gave me leads to many fascinating stories appearing in this work. I thank Dr. Mike Blackford and Mr. Bruce Turner, formerly of the Pacific Tsunami

Warning Center, for explaining the improvements in the warning system. I thank my friend, George Curtis, for his help and encouragement throughout and for getting me started studying tsunamis some fifteen years ago. I am grateful to Drs. Eddie Bernard and Frank Gonzalez of the Pacific Marine Environmental Laboratory, and to Dr. Tom Sokolawski of the West Coast/Alaska Tsunami Warning Center for reviewing portions of this work. A special thanks goes to Dr. Bernard for providing illustrations of the most recent advances in tsunami warning systems.

I am grateful to my friend Bob Chow for insight into the tsunami history of Hilo. I thank Dr. Steven Self for providing me with information on the tsunamis from Krakatau, and Dr. Charles Mader for preparing illustrations showing inundation from tsunamis caused by asteroid impact and for his kind words of encouragement. I am grateful to Emi Takei for translating the commemorative book on the 1896 Sanriku tsunami into English and to Fumio Yamashita for providing excellent illustrations dealing with this tsunami. I thank Dr. Dominique Reymond for showing me around and explaining the operation of the tsunami warning center in Tahiti and Dr. Rajendra Singh for assisting me in learning about tsunamis in Fiji.

I thank the following people, who generously provided me with photographs: Ian Birnie, George Curtis, Lori Dengler, Shaun Fleming, Jeanne Johnston, James Lander, Patricia Lockridge, Chip McCreery, Susan Tissot, Junko Nowaki, Bruce Turner, Gordon Tribble, Dan Walker, and Alexander Malahoff. I am grateful to Amy Cutler for creating the drawings of tsunami generation mechanisms and to John Coney for keeping my computer running and assisting me with preparation of many of the illustrations. Thanks go to Ken Herrick, Kevin Roddy, and Helen Rogers for help in finding promising leads and tracking down obscure documents and to Susan Yugawa and Darin Igawa for their assistance with the illustrations.

I wish to thank my many dear friends in France who took such good care of me while I wrote much of this edition, the Duponchel, Noacco, Terrisse, and Lelièvre families. Special thanks to Germaine Lelièvre for tracking down an important illustration for me.

A special thanks goes to the Pacific Tsunami Museum for encouragement and support in this endeavor, and to my co-author, Min Lee, for cleaning up my grammar, chopping off any dangling participles, and generally keeping me on schedule.

And last but not least I wish to thank my wife, Kamila, for putting up with me, and my children, Malika, Emily, Christopher, and Melanie, who had to forgo many a trip to the beach while Dad was busy writing "the book."

Walter Dudley

It is more than ten years since Walt Dudley suggested we should work together to produce our first book on tsunamis. I had been living in Hawai'i since 1978, and when I started interviewing Hawai'i residents about their tsunami stories I was living in Keaukaha, right at the scene of much of the action in 1946.

In 1985 I returned to Scotland with my family, sorry to leave Hawai'i but taking with me wonderful memories, and particularly grateful that my search for tsunami stories had led me to talk with all kinds of people that I might never have met otherwise. They were all so generous with their time and their often painful memories. Because some of those storytellers are no longer with us, it is especially important that their tales have been preserved. So I thank again, and remember with aloha, all those who helped me with the 1988 edition. Those who told me their stories were Uniko (Yamani) Aoki, Charles (Chick) Auld, Tommy Crabbe, May and Lofty Cook, Robert (Bobby) Fujimoto, Bunji Fujimoto, Takeo Hamamato, Jim Herkes, Kapua Heuer, Fusayo Ito, Bill Jensen, David Kailimi, Frank Kanzaki, Fusai (Tsutsumi) Kasashima, Evelyn Miyashiro, Claude Moore, Lucille and Millard Mundy, Yoshiko (Charlie) Nakaoka, Robert Napeahi, Daniel Nathaniel Jr., Herbert Nishimoto, Hisako Okamoto, Tom Okuyama, Mark Olds, Paul Tallet, Yoshinobu Terada, Josephine Todd, Leslie Waite, Charles Willocks, Masao Uchima, and Mrs. Hideo Yoshiyama.

I have made several return trips to the Big Island. Most recently I spent two months there with the express purpose of talking to more people about their stories. So many more people with stories to tell had been identified, some contacting Walt Dudley, as we had requested at the end of the first book. Others gave interviews to the Tsunami Museum. We decided that a new edition was long overdue, so I was able to combine my pleasure with being back in Hawai'i and seeing the many friends there with the interest of preparing the new book. Susan Tissot of the Tsunami Museum took time from her own work to give me much support and cooperation during this time, giving me access not only to the taped interviews but also to the tsunami essays written by students from Hawaiian elementary schools. She introduced me to Bob Chow—who, as well as giving me his own stories, helped me a great deal with background about the Hilo of times past—and to Cyrus Faryar of KPUA, who showed great interest in the project and copied both the museum tapes and my own interviews. The stories taped for the museum by Charles "Pan" Dahlberg, Wayne Rasmussen, June (Odachi) Shigemasu, Albert Yasuhara, and Marlene (Silva) Young are included in this new edition. From the essays of Keyralee Moses, Thalia Bondaug, Kristin Lee, Michael K. Ishibashi, Malu Debus, Robert Thomas, Elizabeth Rock, Sarah Nonaka, Leina'ala Aina, and Cody Stratton I gleaned the stories of Albino Aguil, Carol Brown, Janet (Kinoshita) Fujimoto, Fusetsu Miyazaki, Edward Medeiros, Matilda Moonie, Carl Rohner, Kimiko Sakai, Everett Spencer, and Keith Stratton.

Again I was most fortunate in finding everyone so generous with their kokua. Many people were ready with suggestions and ideas, and those who had experienced the tsunamis for themselves were willing to give me their time, and in most cases their permission to tape interviews, which now form part of the data at the Tsunami Museum as well as being an important source for this new book. My heartfelt thanks to those whose stories are included here: Peter Alves, Yasuki Arakaki, Carol (Billena) Bufil, Bob "Steamy" Chow, Al Inoue, Jeanne (Branch) Johnston, Bert Kinoshita, Tuk Wah Lee, Wendell Leite, Martha van Gieson McNicoll, Futoshi "Taffy" Okamura, Yasu (Gusukuma) Suguyama, Dr. Tokuso Taniguchi, Lenore van Gieson, and Marlene (Silva) Young.

Finally I would like to thank the two people without whom the book could not have been written—Walter Dudley, who asked me to share this project in the first place, and who has been a good-humored and lively co-author (each of his new chapters a thrill to read)—and my husband, Terry, who has provided very practical help with the electronic demands of modern book-writing (with a co-author thousands of miles away) and maintained an unfailing interest in the book.

Min Lee

Want to Help?

If you are interested in helping to promote tsunami education we can use your assistance. The Pacific Tsunami Museum is in the process of collecting true stories and photographs of tsunamis for use in our education programs and exhibits. We are archiving the scientific papers of tsunami experts from around the world for the museum library. All such donations will be fully acknowledged. The museum can also use financial help to produce, print, and distribute educational materials and to prepare exhibits. All monetary gifts are tax deductible. More information on the Pacific Tsunami Museum can be found at the museum's website:

http:\\www.tsunami.com

or write to

Pacific Tsunami Museum
P.O. Box 806
Hilo, Hawaii 96721.

1

With No Warning

The 1946 Tsunami from the Aleutians

In the early morning hours of April 1, 1946, Officer-in-Charge Anthony Petit and his four-man crew were on duty at the lighthouse in Scotch Cap, Alaska. Located on cold, barren Unimak Island, it is one of the most isolated lighthouses in the United States. The original building, a 45-foot octagonal wooden structure, had been built in 1903 to keep Unimak Pass lighted for mariners using the Bering Sea. In 1940 a new square, white, reinforced concrete lighthouse was built only 50 yards away, rising above the old wooden building to a height of nearly 100 feet. The light's 80,000-candlepower lamp had been moved from the old to the new building, from which it now sent out its powerful 3-second flash at 15-second intervals. The new lighthouse included a fog signal, radio beacon, and on the bluff above and slightly inland, a radio-direction-finding (DF) station, also manned by the U.S. Coast Guard. In spite of the new buildings, it was still a dismal outpost, so difficult to access that the crew would serve for three full years and then be given an entire year's leave. The Scotch Cap Lighthouse, along with the remote Cape Sarichef lighthouse, were known as the "Tombstone Twins" because of the many shipwrecks in this dangerous area of the Aleutian Islands. The name was to be tragically prophetic for the Scotch Cap lighthouse on this April 1—April Fool's Day.

About 90 miles from the lighthouse, the sea floor on the northern slope of the Aleutian Trench began to move, sending powerful shock waves through the earth. Instruments all over the world recorded them; a major

Figure 1.1 U.S. Coast Guard lighthouse at Scotch Cap, Unimak Island, Alaska, prior to the tsunami of April 1, 1946.

earthquake was occurring. The shaking lasted for almost a minute, and when it finally stopped, the Coast Guardsmen on watch at the DF station sent a message to the lighthouse below inquiring of their safety. The crew at the lighthouse reported that no damage had occurred and all were safe. This would be the last communication with the five-man crew of the Scotch Cap lighthouse.

Some 48 minutes later, a 100-foot wall of water struck the Coast Guard installations at Scotch Cap. Water surged over the bluff, reaching 115 feet above sea level and flooding the engine room of the DF station. The commanding officer awakened the off-duty crew and ordered them to higher ground. As they struggled through shallow water, they looked back toward the lighthouse and noticed with foreboding that the light was gone. Not only had the light gone out, but the fog signal and the radio beacon were both silent. The DF radio operator tried repeatedly to raise the light station by both telephone and radio but was greeted with only silence.

Meanwhile in the Hawaiian Islands, at the laboratories of the U.S. Coast and Geodetic Survey on the campus of the University of Hawai'i in Honolulu and at the Hawaiian Volcano Observatory at Kīlauea on Hawai'i, the sensitive earthquake-measuring devices known as seismo-

Figure 1.2 Artist's conception of the giant tsunami wave about to strike Scotch Cap Lighthouse.

graphs had recorded the Aleutian earthquake a few minutes after it had occurred. What no one in Hawai'i knew was that the waves from the Aleutians were spreading out across the Pacific. The Hawaiian Islands lay directly in their path. Though Honolulu was over 2,300 miles from the site of the earthquake, it would take less than 5 hours for the waves to reach Hawai'i.

At first light in Alaska, the Coast Guardsmen at Scotch Cap descended to the edge of the cliff above the lighthouse. As they peered down, they saw with horror that the lighthouse was gone—both the old wooden building and the new concrete lighthouse had been completely washed away. There was no trace of the five-man crew, no bodies, no clothing—nothing. The DF station immediately sent the following message to other Coast Guard installations:

> TIDAL WAVE PRECEDED BY EARTHQUAKE COMPLETELY DESTROYED SCOTCH CAP LIGHT STATION WITH LOSS OF ALL HANDS X TOP OF WAVE STRUCK THIS UNIT CAUSING EXTENSIVE DAMAGE BUT NO LOSS OF LIFE X ENGINE ROOM FLOODED OUT BUT EXPECT BE ABLE MAINTAIN EMERGENCY POWER IF NO FURTHER DAMAGE EXPERIENCED X WILL SEARCH FOR BODIES OF SCOTCH CAP PERSONNEL AS SOON AS POSSIBLE X REQUEST INSTRUCTION AND ASSISTANCE.

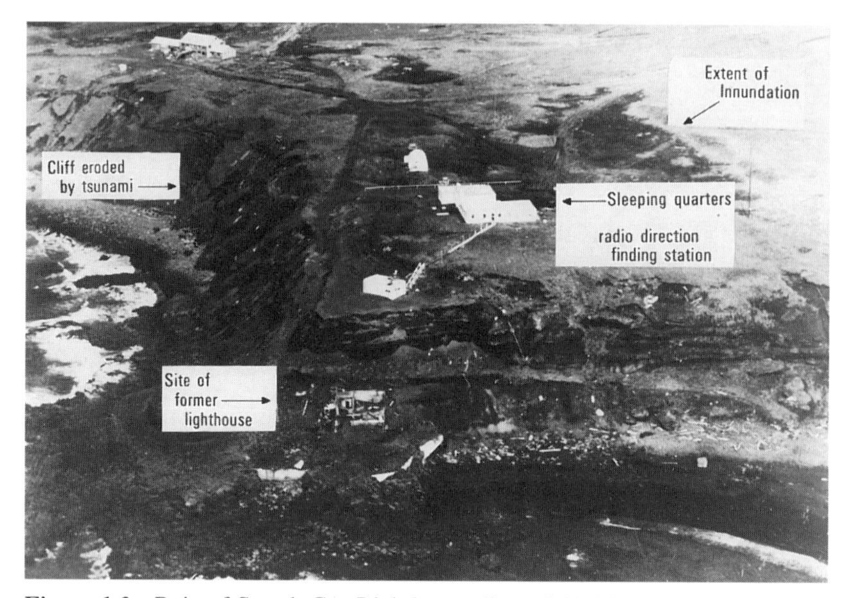

Figure 1.3 Ruin of Scotch Cap Lighthouse *(lower left)* following the tsunami of April 1. Radio-direction-finding station and crew quarters shown on the bluff above were flooded *(center)*.

The tsunami of April 1, 1946, had claimed it first victims. They would not be the last. The first wave reached Kaua'i at 5:55 A.M. and Honolulu on the island of O'ahu just after 6:30 A.M. It would reach the island of Hawai'i just before 7 A.M.

On the south shore of Kaua'i, work had already started at the survey office of McBryde Sugar Company. The supervisor told Fusetsu Miyazaki and his fellow workers that he had heard that a tsunami had begun to strike the island and they could go and look from the cliff top. They all went to see, thinking that maybe it was an April Fool joke. When they reached the cliff top they were amazed to find that there was no water in the bay. Everywhere fish were jumping and flopping back to the sand and mud. As they stood and stared, water started to flow back into the bay. When it reached the shore it did not stop, but kept flowing up the valley, snapping off tree branches as it rose higher and higher. Then the bay emptied again. Mr. Miyazaki and his friends climbed down to examine the debris on the floor of the bay, the fish entangled in tree branches. They were wise enough to keep an eye on the ocean, and when the water started to advance again, they scrambled quickly up the cliff. They were lucky on this April Fool's Day. Their curiosity had not proved fatal.

To the west of the island of O'ahu, the destroyer-minesweeper U.S.S. *Thompson* was headed toward Pearl Harbor. *Thompson* was returning from Bikini, where the vessel had been involved in preparing the atoll for the forthcoming A-bomb test. Seaman Perry Minton, a former resident of Honolulu, had gotten up hours before dawn and was in the ship's radio room waiting for the island of O'ahu to show up on the radar screen. The radar man asked Perry to fill in for him while he ate breakfast. "Almost as soon as I put on the headset, I heard a patrol plane—probably a PBY—calling its base at Kaneohe to report something on the surface of the sea, perhaps just a line or small wave." When the station at Kāne'ohe asked the pilot to drop down closer to the surface in order to identify the phenomenon, the pilot said it was gone—"it had outrun his aircraft." Minton wondered what was going on so far to the north of O'ahu, but his friends aboard the *Thompson* assured him that it was just an "April Fool's joke."

Not long afterward, "the radio frequencies were filled with messages to and from ships in Pearl Harbor, many of which had been moved by an abrupt surge. *Thompson* was told to stay offshore and out of Pearl Harbor until things were unscrambled.

On the north shore of O'ahu Dr. Francis Sheppard and his wife were staying in a cottage on the beach. Sheppard, a distinguished marine geologist with Scripps Institute of Oceanography, gave the following account:[1]

> [W]e were sleeping peacefully when we were awakened by a loud hissing sound, which sounded for all the world as if dozens of locomotives were blowing off steam directly outside the house. Puzzled, we jumped up and rushed to the front window. Where there had been a beach previously, we saw nothing but boiling water, which was sweeping over the ten-foot top of the beach ridge and coming directly at the house.

After witnessing the arrival of the first wave, Shepard got his camera, left the house, and saw the water retreating until coral reef was exposed and stranded fish were flapping.

---

[1] *The Earth beneath the Sea* (rev. ed., Baltimore: Johns Hopkins University Press, 1995), pp. 31–34.

> Trying to show my erudition, I said to my wife, "There will be another wave, but it won't be as exciting as the one that awakened us.". . . Was I mistaken? In a few minutes as I stood at the edge of the beach ridge in front of the house, I could see the water beginning to rise and swell up around the outer edges of the exposed reef; it built higher and higher and then came racing forward with amazing velocity. . . . As it piled up in front of me, I began to wonder whether this wave was really going to be smaller than the preceding one. I called to my wife to run to the back of the house for protection, but she had already started, and I followed her just in time. As I looked back I saw the water surging over the spot where I had been standing a moment before. Suddenly we heard a terrible smashing of glass at the front of the house. The refrigerator passed us on the left side moving upright out into the cane field. On the right came a wall of water sweeping down the road. We were startled to see that there was nothing but kindling wood left of what had been the nearby house to the east.

As the wave subsided, the Shepards hurried to higher ground, just ahead of a third and still larger wave.

> We started running along the emerging beach ridge in the only direction in which we could get to the slightly elevated main road. . . . As we hurried through this break, another huge wave came rolling in over the reef and broke with shuddering force against the small escarpment at the top of the beach. Then, rising as a monstrous wall of water, it swept on after us, flattening the cane field with a terrifying sound. We reached the comparative safety of the elevated road just ahead of the wave. . . . Finally, after about six waves had moved in, each one apparently getting progressively weaker, I decided I had better go back and see what I could rescue from what was left of the house. . . . I had just reached the door when I became conscious that a very powerful mass of water was bearing down on the place. . . . I rushed to a nearby tree and climbed it as fast as possible and then hung on for dear life as I swayed back and forth under the impact of the wave.

Laupāhoehoe

It was just another Monday morning in paradise for Frank Kanzaki, a young teacher at Laupāhoehoe school on the Hāmākua coast of the island of Hawai'i. As he strolled from his oceanside cottage to that of his friend Peter Nakano, he looked around at the familiar scene and breathed in the early morning air. It was 10 minutes before 7 A.M. The surf was high, breaking on the *pāhoehoe* lava rocks that give the small peninsula its name. A flat expanse of grass behind the seawall provided a baseball field, where students from the early buses were playing and talking. Every

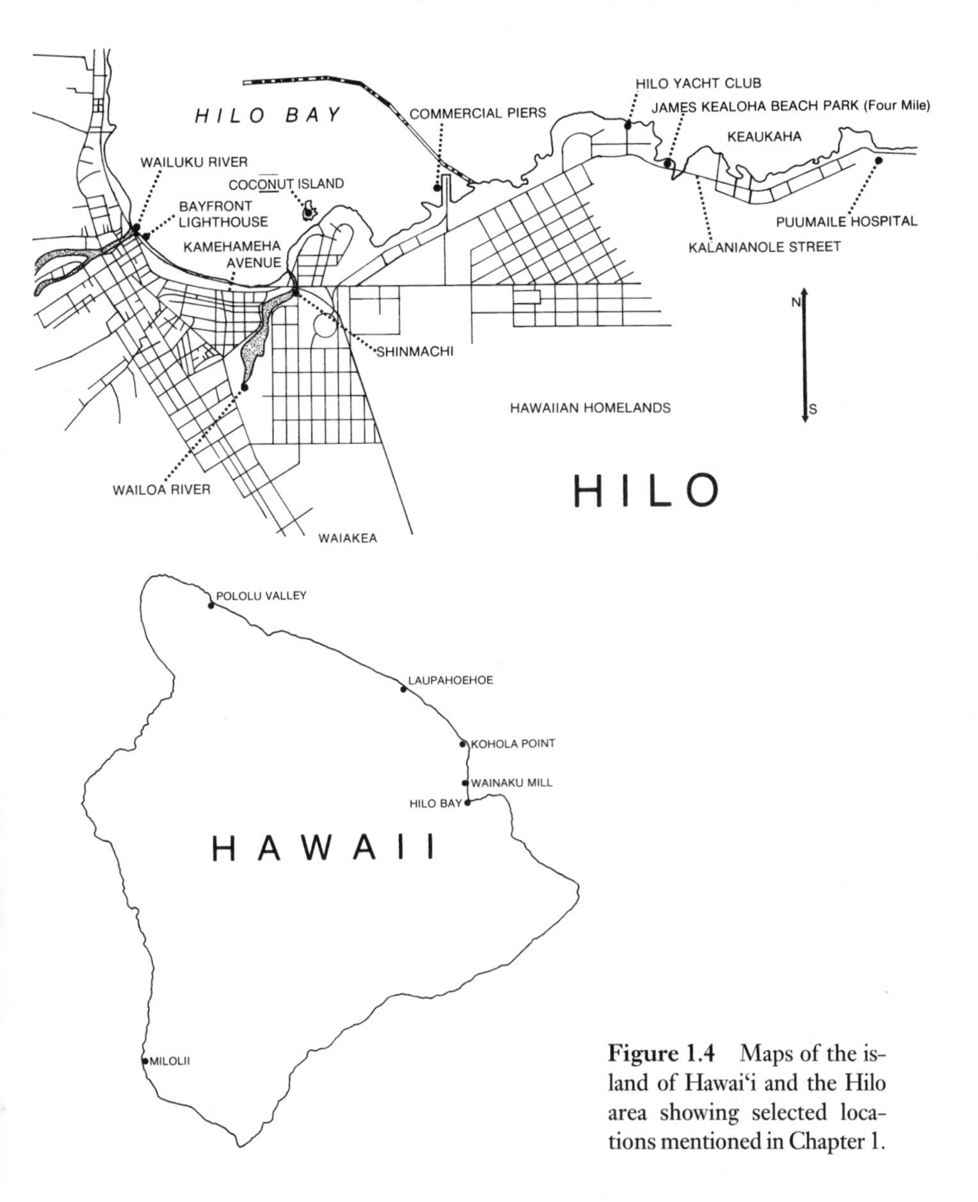

Figure 1.4 Maps of the island of Hawai'i and the Hilo area showing selected locations mentioned in Chapter 1.

morning Frank took breakfast with the Nakanos. He smiled as he entered and took his place at the table between the two girls, Christine and Stella, and opposite Peter and his wife and baby. Through the window he could see the ocean—the waves a little higher now—and some of the students on the seawall and on the shore.

At that moment the day ceased to be like any other day. As Frank glanced out at his carefree pupils, he saw behind them a wave that seemed at first big, then gigantic. Jumping up, he seized Christine and Stella and pushed them to the back of the house, toward the cliff. As he reached the back door, the water shattered it. The wave swept away the stilts on which the house stood and carried the cottage for a few seconds. Then the building began to disintegrate. The ceiling collapsed. Frank found himself struggling in the water and knew that he could not stay afloat with both arms encumbered. He opened one hand—Stella was carried away. Now that he had one arm free, Frank was able to support Christine and keep his head out of the water. As the sea ebbed back to its shore, he discovered that one foot could touch firm ground. He braced himself and was not pulled out to sea by the receding wave. Frank was aware of Christine next to him. He raised his head and saw Peter nearby, his arm trapped between a tree and a rock. Of Peter's wife and baby there was no sign. Frank's gaze swept the area. In a space of ten minutes the scene had changed completely and forever. The wave had deposited him at the far side of the baseball field; along its edge were smashed cottages, lumber, tree branches, and the collapsed grandstand. Among the wreckage, people began to stir and call out. In the ocean were many heads, bobbing in the heaving sea.

Many of the people washed out to sea were students who had been playing near the seawall or on the shore when the wave struck. Like Frank, they had thought this just another Monday, thought that the only excitement would be an April Fool trick.

On the early school bus that carried Bunji Fujimoto to school that morning, the students had been well aware of the fact that it was April Fool's Day. Laughter of disbelief greeted the information volunteered by some, as the bus came down the hill, that the sea had receded, leaving the ocean

Figure 1.5 Laupāhoehoe Point prior to the tsunami of April 1, 1946. Note the playing fields and cottages nearest the ocean.

floor bare. However, when the doubters were finally persuaded to look from the bus window, they discovered it was no joke. Excited by this unusual event, but completely unaware of its significance, they left the bus as soon as it stopped and ran to the seawall. Others, a little less curious, or more cautious, stayed on the grass. Bunji was one of those who kept their distance from the ocean. As he watched his more daring classmates—one of them his brother—he saw the water rise over the seawall. At first Bunji did not realize the danger. It was not a crashing, breaking wave, but an ever-rising, ever-encroaching wall of water that flowed without stopping. Scared now, he turned and ran across the grass toward higher ground. Other students fled with him, and he noticed particularly a member of the basketball team who was making good use of his long legs; he had been on the ocean side of the seawall when the wave began to surge and, with his natural speed enhanced by terror, he achieved safety. Running too was Yasu Gusukuma, a sixteen-year-old who had been on the same school bus as Bunji. Like him, Yasu had been too nervous to go down to the shore and look at all the fish flopping on the bare sea bed. Instead, she stood (where the tsunami monument now stands) with her friend and watched two waves come in, and then she saw that the third was bigger. Suddenly the water was up to her ankles. She began to run away from the sea. After she had crossed the ballfield, she ran up some concrete steps, picking up a little girl as she did so and setting her down when she reached higher ground. In the confusion and terror that followed, she lost contact with the child and with her friend. Gripped by a horror that would remain forever in her memory, she saw water coming from all sides of the point and "boiling" in the center. She saw cottages spinning on top of the water, she saw the grandstand collapse, then she turned and ran up the hill as fast as she could.

Up the hill was where Carol Billena had been urged by her older sister, Ascension. Carol and her sister, second and sixth graders, normally walked to school from their home on the eastern flank of Mauna Kea. There was no bus because the house was not in a plantation camp. This particular morning it had been raining hard on the slopes, so hard indeed that her father was unable to do his work on the sugar plantation. He decided to drive into Hilo instead and gave them a ride to school. This meant that they arrived at school very early, each girl going to sit in front of her own classroom with classmates. Carol watched the school buses arrive and saw some of the children, excited and eager, running toward the shore. Intrigued, she called out to ask what was happening; one of the running boys shouted at her, "Tiger wave!" Even more intrigued, she pictured a big tiger, and wanted to see for herself, but she thought she had better tell her big sister first. She dashed to Ascension's classroom and announced that she wanted to go with all the other kids and see the "tiger wave," and was startled by the reaction. Her sister sprang up, dropping all her things, and pushed Carol out of the classroom. "Come on, it's a *tidal* wave. Run!" They ran as fast as they could, Ascension repeating urgently, "The wave is coming, the wave is coming. Run, run, run!" When they got past the Japanese school on the hill, they turned to look and were confronted by a horrifying scene. As with Yasu, the image that came to Carol, and stayed with her, was one of "boiling," as the water surged under the cottages. As they looked at the lauhala branches, roofs, and boards jumbled together in the water, Carol thought about the baby in the cottages to whom she had taken a gift from her teacher just days before.

There were children outside the cottages as well as people inside them—students who had run to look at the strange sea. Marlene Silva lived in Laupāhoehoe near the hospital; she and her sister were given a ride down the old road from the highway by their aunt, the cafeteria manager, known to all as Miss Alexandrina. From the road they could see, as those on the school bus had, that the sea had withdrawn from the bay and the sea floor was exposed. But none of them had any idea of what this meant, so they continued down to the school. Marlene, nine years old and a third grader, went to her classroom to put down her books and met her friend Janet De Caires. Another classmate mischievously told her, "If you go down to the landing, you'll cry, because the ocean is getting wider." Curious, she went with Janet to ask Miss Alexandrina to take them to look at this strange phenomenon, but her aunt said she was too busy. She told them to go instead with one of the Ugaldi boys from the high school. They walked toward the tennis courts to get a look at what was happening. The first wave they saw demolished the old boathouse at the landing but did no other damage, so Marlene still wanted to go down to the sea for a better look, fascinated by the sight of fish and turtles that had been washed up by the boys' bathroom. But in spite of her curiosity, she was afraid. Janet caught sight of her brother John and her sister Madeleine walking near the cottages, and she wanted to join them. "Come on," she urged, but Marlene felt an even stronger fear and said, "Don't go, stay with me." Jane wanted the security of being with her older siblings, and ran toward them. As Marlene watched, and as Janet reached the middle of the field, the water came from all directions, as high as the coconut trees, boiling and moving fast, filled with debris. She turned and ran past the grandstand, where she glimpsed children on the bleachers as the wave smashed into the structure, on up to the girls' bathroom. After seeing the wave cut the boys' bathroom in half "like a knife," Marlene did not look around again but ran as fast as she could, surrounded by other fleeing children, some much bigger than she was, so that she was in a forest of legs.

Some of the panicky students were overwhelmed by the wave but not dragged out to sea because they became stuck in the bushes. Others, like Bunji's younger brother, were swallowed by the rushing sea and never seen again. Still others were blessed with a special fortune: although they had been tossed into the sea, they lived and would remember the experience with awe.

Herbert Nishimoto would never forget that April Fool's Day. A tenth grader, he lived at Ninole but had spent the weekend "camping" with a fellow student, Mamoru Ishizu, in the cottage that Frank Kanzaki shared with Fred Cruz. Herbert was alerted to the situation by the cries of a friend, Daniel Akiona, who ran past the cottages shouting, "Tidal wave!" Daniel's family lived in a house on the point opposite the boat ramp. He and Herbert watched the waves come and go, and they saw the second demolish an old canoe shed by the shore. Then Daniel urged that they go to his house. The third wave took longer to gather than the other two had, but it looked terrifyingly huge. Desperate for refuge, they dashed toward Daniel's home, but as they reached the door the whole building collapsed. Herbert was lifted by the wave, sucked into the ocean, and deposited on the reef. As he gasped for breath and stared around him he saw his friend Mamoru floating on a log and a teacher floundering in the water. Herbert tried to remove his blue jeans, but they were so tight around the ankle that he had managed to remove only one trouser leg when another wave arrived. As he dove under the wave the heavy pants, trailing from one foot, caught on the reef, and he was battered to and fro. When his confusion cleared he found himself floating amidst debris, and accompanied by sharks. Fortunately, his good luck did not desert him. A section of flooring from one of the cottages floated nearby; he heaved himself onto it, and lay wondering what would happen next.

Herbert's feeling of stunned shock and fearful anticipation were shared at that moment all along the island's eastern coast. The April morning had turned from unremarkable to unbelievable, and nowhere more so than around Hilo Bay.

Honoli'i

Down at Honoli'i cove Wendell Leite had been preparing for school. Seventeen years old, he lived in a house just 10 feet above the high tide mark with his four sisters and one brother, his parents, and his Hawaiian grandmother. They all went out into the yard to observe the strange behavior of Honoli'i Stream: the water from it was disappearing into the ocean. None of them had ever seen such a thing before. As they stood and stared, they heard a voice hailing them from high above on the Honoli'i bridge. It was George Meyers, a driver for Hilo Transportation and Terminal (HT&T). "Get out, get out, it's a tidal wave!" he yelled. So they piled into their cars, saving nothing from the house, and drove uphill as quickly as possible. When they stopped to look back they saw a huge, dark mountain of water that seemed to build up and then flow out in all directions. It filled the lower part of the valley, destroying the ground floor of their home, including his grandmother's bedroom with all its treasured possessions. Sad as Wendell was to see his grandmother's heirlooms and his parents' hard work demolished by the wave, and to lose his new trombone, he knew that George Meyer's warning had saved their lives, and that they had been very lucky to have a quick route to high ground.

Hilo: The Bayfront Area

From their home on Pukihae Street, atop a 30-foot-high sea cliff overlooking Hilo Bay, Kapua Heuer and her family saw the events unreel before them. Kapua was busy preparing for the day's activities when one of her daughters asked, "Mommy, what's wrong with the water?" They all went to the cliff's edge at the end of the yard and saw that the seafloor was becoming exposed as the water withdrew. Far out at the breakwater the outward flow met an incoming wave, and the whole mass of water rushed toward the shore. Instinctively, Kapua and her daughters stepped back—just in time. As the wave collided with the sea cliff, water splashed over the tops of their coconut trees. Then the crash of the arriving wave mingled with the sound of walls and buildings being crushed as the wave struck downtown Hilo. The ocean was filled with debris and with people struggling in the waves that continued to flood into the bay. At each retreat of the water more flotsam filled the bay accompanied by a loud sucking sound, as if the ocean drank Hilo's offerings. Safe atop the sea cliff but unable to help, Kapua watched the struggles of the victims and the destruction of the bayfront.

On board the S.S. *Brigham Victory*, First Mate Edwin B. Eastman was all too aware of their danger. Theirs was the only ship in the harbor, having had to wait out in the bay for a berth until the previous day. The ship's cargo included 50 tons of dynamite and the volatile blasting caps that could set it all off. On deck was a load of lumber from which the lashings had been removed in preparation for its unloading. The captain was staying on shore, so the first mate was in charge. Thinking that the lines would hold them, Eastman had not been too concerned when the water first began to drain from the harbor. After the first wave, which broke three of the hawsers, he changed his mind and ordered the engines to be started. It took only seven minutes for them to achieve full power, so by the time the biggest wave hit they were able to use that power to help keep out of trouble. Also on board was one of the stewards, Wayne Rasmussen. He had noticed movement on the vessel when he was setting up for breakfast, went up onto the weather deck and saw the first wave, and decided to go below for his camera. As the ship maneuvered, he and his shipmates heard the clash of the corrugated asbestos sheets of the warehouse wall as they were torn away, and they witnessed with horror a stevedore on the pier being engulfed by the monstrous wave. Wayne recorded the scene in amazing photographs. The harbor was filled with debris, including railroad cars. The ship finally managed to avoid the reefs in the very shallow water in the partially drained bay and get past the breakwater to the open sea. Before doing so they picked up some survivors, one of them a truck driver whose truck had been washed into the bay. They threw a line to him as he floated by on wreckage and were able to take him aboard. He was one of three drivers for the Kaʻū Sugar Company in Pahala who had been unloading sugar on the dock when the waves struck.

Another man saved by the ship had a miraculous escape that would influence his whole life. Tuk Wah Lee was a stevedore at Hilo docks, twenty-seven years old and ready for a day's work unloading the *Brigham Victory.* Having arrived at work at 6:30 A.M., he was "talking story" in the warehouse with some of his co-workers when someone outside yelled that the water had gone down and they could see the sea floor. Everyone rushed outside to look, and several of the men went to pick up fish. The water then rose with great force, whistling through the cracks in the concrete dock. The assistant superintendent hurried over to Tuk and the others and asked for help in securing the ship, as some of the mooring lines had snapped. After that had been done, Tuk wanted a better look, and decided to climb up an old Coast Guard tower at the end of the dock. He had taken only a few steps up the ladder when he saw a brown wall of water advancing across the harbor. What scared him was the fact that as it got nearer, it got bigger. He yelled a warning to the general superintendent of the sugar warehouse, then ran into the warehouse himself through the north door. He scrambled up a ladder attached to one of the supporting pillars and up onto a walkway that ran along the rafters. The noise that Wayne Rasmussen had heard, of the walls crashing and clashing as the wave struck them, sounded deafeningly loud to Tuk. The walls nearest the ocean were swept away. Moving along the walkway in the direction of the shore he looked down and saw that the 25-foot-high piles of sugar had crumbled and that boxcars on the railway track had overturned. A barge had been washed right through the warehouse, banging its way through the west wall and out the other side. But the water was receding. This made him very happy, and in thanks he took off his shoes and said a short prayer. After that he felt a confidence that he would be all right, and even felt that "the place had a nice glow." In this frame of mind he decided it was time to climb down from the rafters. When he ran out onto the dock he decided that the safest place would be on board the ship. He called to the crew to throw him a line, but when they failed to do so immediately he felt that he had to swim to the ship, which was floating free about 40 feet from the dock. The water was filled with debris, but he dived in and swam rapidly to the ship's side, where a gangway was dangling. He climbed up it to the ship. All this had taken him about four minutes. Just as he was hauled on board another wave arrived to carry away the gangplank—and also to flatten the warehouse. Either he was a very lucky man indeed—or he had been saved for a purpose.

Figure 1.6 Stevedore on the commercial pier in Hilo, victim of the largest wave to strike Hilo harbor.

Figure 1.7 The scene shown in Figure 1.6 as the next wave submerged the wharf.

Tuk Wah Lee's experience on this day contributed to his decision to become a minister. But at that moment he was simply relieved and thankful to be on the ship that had saved him. It headed for Maui, after having been part of the horrifying scene presented to those shocked spectators who were beyond the reach of the waves.

From the vantage point of the highway bridge on the Wailuku River, Jim Herkes had his own perspective on the event. Just home from the army, he had been giving his brother Bob a ride from their home to Hilo Intermediate School. As they drove over the bridge they noticed the exposed river bottom, and Jim was sure that a tidal wave was imminent. Fascinated by the thought, he parked his car and went back to the bridge to watch. Like Kapua on her cliff top, Jim and Bob watched in amazement as the harbor drained. When the first big wave swept in, a span of the railroad bridge was carried upriver. As the water receded, the span was pulled back toward the ocean: then it swept beneath them and was deposited on the small island in the Wailuku known as Maui's Canoe. Jim could hardly believe what he had seen from this vantage point only just above the churning water.

Daniel Nathaniel Jr. (known as Baby Dan to his friends) soon discovered that the river's edge was not the place to be on this April morning. Sometime earlier his cousin Alona had called to say there would be a tidal wave—she had seen the water recede from the bay. "April Fool" was his first reaction—like that of so many others. He decided, however, to see for himself, and went to the mouth of the Wailuku. He had seen small tidal waves before, yet realized with a jolt that this was to be a very different experience. As he stood on Shipman Street, a huge wave roared into the river, surged over its banks, and lifted him several feet into the air. As it carried him forward he clutched at one of the roof beams on the Amfac building, and was left dangling as the water receded. Dan was shaken but safe. His slippers were in the ocean, he was in the air, but he was alive and kicking!

Figure 1.8 Series of photographs of a wave progressing up the Wailuku River in Hilo. Note the span torn from the railroad bridge and deposited in the river *(top photo)*.

Fred Naylor would not be so lucky. The eighty-four-year-old New Zealander, a well-known local character with a red nose and convivial habits, was at the Hilo railroad station when the waves arrived. In an act of heroism, he pushed a young girl, Harriet Kama, into the station house and slammed the door just as the wave reached him. Then he was gone, lost to the sea.

All were taken unawares; there was absolutely no warning. Some were in the wrong place at the wrong time, while others escaped without knowing until later that they had been in danger.

Bill Jensen drove into town from the Wainaku mill of Hilo Sugar Co. (a mill just north of Hilo at that time) and saw that the railroad bridge had been swept away. In his ignorance, an ignorance shared by the vast majority of Hilo's population, he did not know that there would be more waves to come. He drove along Kamehameha Avenue, exposed to the ocean, but fortune smiled on him: because his path was blocked by lumber washed from a local lumber yard, he turned uphill, away from the bayfront. Utterly oblivious to the action of the ocean, he did not discover until later that just as he reached Keawe Street, a wave rolled over Kamehameha Avenue and would have claimed him as another victim had it not been for the pile of lumber in the street.

Bill had been unaware of the imminent wave when he drove south along Kamehameha. Albert Yasuhara, however, saw the start of its destructive arrival and chose to put himself directly in its path. A salesman for the Coca-Cola company, he was driving north along Kamehameha Avenue to his Hāmākua route (up the east coast of the island), with a load of 202 cases of soda, when he saw a man running fast from the direction of Moʻoheau Park, looking scared. For a moment he wondered who was chasing him. Then he saw a wave halfway across the park. He stepped on the gas, turned left into Ponohawai Street and parked at what seemed a safe distance, uphill of Kinoʻole Street. But his reaction was not to heave a sigh of relief and go about his business. He guessed it was a "tidal wave" and had heard that there were at least three waves involved in such an event—so he went back to wait for the next one. After waiting for a while, he decided that nothing else was going to happen and began to walk uphill toward his truck. Suddenly he heard a roaring sound, and buildings crashing—but he did not run away. He ran *toward* the sound and arrived on Front Street to see that all the buildings on the ocean side of Kamehameha Avenue had been pushed by the second wave into the middle of the road. Possessed of a dangerous curiosity, he was determined to see what would happen if another wave came, and waited. First he saw the bay empty, the sea floor exposed, the water piled up on the ocean side of the breakwater. Then the wave overwhelmed the breakwater and rushed towards the land. When it hit the shoreline Albert saw that the water reached halfway up the coconut trees, and knew that he was definitely in the wrong place. "Here it comes!" he yelled, and ran up Mamo Street with the other people who had been watching, just ahead of the advancing water.

At around this time, on Ponohawai Street, Taffy Okamura and his coworker were running *in* the water, hoping frantically that it would not overtake them. They had been working on the bayfront unloading goods from the railroad into the T. H. Davies warehouse, when the wave action started. One wave broke against the step, then they saw the water recede from the bay, and watched in curiosity, with no inkling of their danger. But the next wave was bigger and moved the boxcars, so now Taffy was

Figure 1.9 Men who had been working along the Hilo bay front running from the third and largest wave.

frightened, and said "Let's go." They were still on Kamehameha Avenue when the water flooded over. They ran up Ponohawai Street for all they were worth, on to Keawe, and along to Wainuenue, where they saw the school bus stranded on the corner of Kamehameha Avenue.

There were many people running on Ponohawai Street. Peter Alves, not long arrived in Hawai'i after his military service in North Africa and Sicily, had been standing in front of the shop at the Mechanics School on Kamehameha Avenue when he saw the big wave hit the shore, and he took off uphill with all speed. Like most of the other fugitives from the wave, he had had no previous warning.

Bill Jensen, Albert Yasuhara, Taffy Okamura, and Peter Alves had all seen the destruction wrought by the previous wave—the businesses and shops from the ocean side of Kamehameha Avenue heaped into matchsticks on the other side. On Mamo Street, in the second floor apartment over the Okamoto store, Hisako Okamoto had looked out of the window just in time to see the wave tower above the downtown bandstand. But her family was among the more fortunate—they suffered financial loss, but not personal tragedy.

Figure 1.10 Waves washing up Waiānuenue Avenue in Hilo.

Hilo: The Shinmachi Community

The greatest destruction and loss of life occurred on the coastal strip running along the bayfront and to the north of the Wailoa River. In the area that is now Wailoa State Park, there existed a close-packed, close-knit community known as Shinmachi. Founded at the beginning of this century by Japanese immigrants (*Shinmachi* in Japanese means "new town"), it was bustling on April 1 with the same activities that were occupying other islanders—getting ready for school and work.

In the Tsutsumi household, next to the Wailoa River, fifteen-year-old Fusai looked up in surprise as her sister rushed in saying, "Come see the river! It goes so fast!" Together the two girls went to look and were confronted by several boys pedaling their bicycles as fast as they could go. "Tidal wave! Tidal wave!" The sisters ran inside to warn their brother. He laughed at their simplicity, "They trick you—April Fool!" In that very instant the wave arrived with a symphony of crashes. Through the window they could see the river full of people and debris, all mingled together as the water flowed back to the ocean. They heard their neighbor, Mr. Sakido, calling to them to go to the Coca-Cola bottling plant, a two-story concrete building. They ran through his house and took refuge on the second floor of the factory with many other Shinmachi residents. All of these people were to owe their lives to the fact that the factory manager had arrived early that day and the door was unlocked.

Fusai's brother Hiromo had decided to run back to their house to see what he could save. The adults prevented the younger people from leaving the Coca-Cola building, but Hiromo was old enough to make his own decisions. Once back at his house, however, he could not decide in the excitement of the moment what to take away with him. Looking around, he caught sight of a *furoshiki* (a cloth wrapper used for carrying small articles) full of family photographs and pictures. He seized it and ran back to the safety of the Coca-Cola building. The next day his father protested that he had not saved the considerable amount of money in the house. But in the years to come the mementos were of far greater value than mere money. Many survivors echoed the Tsutsumis' feelings: the loss of personal effects was of more importance in the long term than the destruction of buildings or the loss of money. At the time, however, those in the Coca-Cola building were thinking only of survival.

Masao Uchima was there, too. He had been in bed when he heard his father shout in Japanese, "Tsunami!" Water was everywhere; it seemed to him that the island was sinking. His father shouted for everyone to go to the Coca-Cola building—they reached it before the next wave arrived. Thus Masao became an observer, not a victim. From the roof of the building he saw a 15-foot wave roll in, followed by an enormous wall of water that stretched right across the bay. This wave obliterated most of Shinmachi, except the building where he stood.

The Coca-Cola building provided shelter also for the Inoue family. Four-year-old Al had been asleep in the family apartment over the meat market and grocery store; then his mother began shaking him and telling him they had to leave. He ran first to the window overlooking Kamehameha Avenue and was amazed to see a boxcar lying on its side next to the railroad tracks. Then his cousin picked him up and carried him through waist-high water to the Coca-Cola building. He remembers that many of the adults were gathered around an old Japanese man, who was on his knees and praying. But there were many people praying all over Hilo. Outside the building, in the river, those who had been swept away by the wave prayed where it had left them.

It was after the second wave that Yoshinobu Terada found himself in the river. When the water began to recede the first time, his father had sent him to look at their boat. As he was returning home, he saw a 4-foot wall of water rolling toward him up an alleyway. He leapt onto the nearest porch. From all around came the crash of structures crumbling—a row of houses telescoped and the buildings rammed together. The house on whose porch he was standing was carried into the river and began to sink. He jumped into the water, where there were many other people and a vast amount of debris; he heard the cries of people struck by swirling beams, appliances, doors, and roofs. All was turmoil and time stood still. Fortunately, however, Yoshinobu focused his thoughts on the sight of his brother's surfboard floating by him in the shambles. He climbed onto it, happy to be raised above the wreckage with a means of reaching shore. He paddled to land just before the next wave arrived, and he went to look for his family in ravaged Shinmachi.

At the southern edge of the Shinmachi township stood the Hilo Iron Works. Edward Medeiros had arrived at his machine at five minutes before seven. When he looked out through the open gateways he was amazed to see the water receding in the bay. Then the water started to pile up, advancing toward the shore. With fear and horror he realized it must be a tidal wave, but he was unable to shout. Forcing his legs to move, he ran to a crane and scrambled up it, clinging tightly as the wave's force shook the building and tumbled the steel rollers as if they were made of paper. When the water had receded somewhat, he ran from that crane to a higher one, and from it looked out over Shinmachi. He could hardly believe what he saw. The township was smashed, the buildings scattered and demolished. Some were in the Wailoa River, a river now crammed with debris, and with people. In the water and in the trees along its bank he saw people crying and screaming, their clothes ripped from them by the force of the wave.

Still inside one of the houses he saw in the river was eleven-year-old June Odachi, with her mother, three sisters, and one brother. They lived in a chapel near the iron works. Like other children in Shinmachi, June had been preparing for school when she heard a big commotion. The family had no time to take action before the wave crashed into the building, lifted it from its foundations, and carried it up the river.

The Odachis were in their house, and still together, but many others were alone and in the water. Ten-year-old Matilda Moonie had been getting ready for school in the apartment where she lived with her mother, four brothers, one sister, two uncles, and an aunt. The first wave shook the building and scared Matilda, so she ran to find her two younger brothers. When the third wave struck, the building collapsed and she was tumbled in the water with the debris, in darkness and terror. Her brothers had been torn from her arms, and when she got her head above water and looked around she could not see them. She did see her grandma, her sister, and her uncle being swept up the river. She was traveling the other way, behind Hilo Iron Works. She climbed onto a broken roof, but another wave knocked her off. She sank to the bottom of the river and struggled back onto wreckage, praying for help from God. Her prayers were answered in the form of Rango Inoue, Al's father, who left his refuge in the Coca-Cola bottling plant when he saw her in the debris, pulled her free from the flotsam, and took her back to the building. Like many others, she had been stripped of her clothes by the waves, so they dressed her in a burlap bag with holes cut for her neck and arms. Though bruised, confused, and worried for her family, she was alive, and had escaped from the river.

On the other side of that river was Waiākea town. Here Hilo had begun its commercial expansion with the building of the railway terminus in 1899.

Waiākea

In 1946 there was a gas station and store at the crossroads next to the bridge, as well as a newly opened café. Its owners, Evelyn and Richard Miyashiro, had started their business in January. Now they were on the bridge with the other Waiākea residents, waiting to see the big waves. There had been other tidal waves before, when the water had risen only a few feet. With the arrival of the first two waves, excitement rose among the watchers. The third wave, however, was immense—there was no time to run. The Miyashiros were there when the massive wave swept over the bridge, and, amazingly, they were still there after it had receded. Evelyn Miyashiro, eight months pregnant with her first child, clung to the concrete uprights of the bridge, her husband beside her. Coconut trees were flattened, the building housing the Kitagawa auto dealership collapsed, and the Miyashiros' café was flooded to a depth of 3 feet—but Evelyn and Richard were still alive.

Unlike the Miyashiros, the Hideo Yoshiyama family suffered the loss of several loved ones. Unaware of their peril, Mr. and Mrs. Yoshiyama and their young son Alan went outside their home on the bank of the Wailoa River to watch the waves arrive. After the first waves, several people began shouting that a "big one" was coming. Mrs. Yoshiyama returned to the house to warn her parents, Mr. and Mrs. Fukui. Meanwhile, the next wave arrived with great force. Hideo was swept by the wave into the river—his son Alan torn from his grasp.

Inside the house, Mrs. Yoshiyama was having a difficult time convincing her parents to leave the building—her father was reluctant to leave his plants. Mrs. Yoshiyama gave up the attempt to persuade him and ran upstairs to the living room with her mother and her niece Alison. The wave invaded the house and carried them across the room in a jumble of lumber and furniture, Mrs. Fukui clinging to her daughter with one hand and clasping her grandchild desperately in the other. The powerful water pulled at her as it moved back toward the ocean. Her daughter was stuck fast in the debris, but the inexorable pull of the flowing water must have made Mrs. Fukui fear that they would all be swallowed in the muddy torrent. She snatched her hand free. Then she and Alison slipped down the bank and disappeared into the sea.

Mrs. Yoshiyama still mourns her three-year-old son and her mother and niece, but feels fortunate that her two daughters, who had been on their way to school, escaped without injury. Hideo, her husband, was spared as well. Her father was swept from the basement room where he had remained with his plants but was washed up alive on the bank of the river.

Some residents of the Waiākea district were more fortunate than others. When Mrs. Josephine Todd looked out into the bay that morning and saw the water recede, she was one of the few who thought she knew what was going to happen. She sent her children with her niece up to her sister's home on Kaumana Drive. Meanwhile, she busied herself moving possessions from her home to her niece's house, which was nearby but on higher ground. Her idea was to save them from a soaking. Like others in Waiākea she had seen tidal waves before, but only small ones. Familiar with the unpredictability of the ocean, she kept a watchful eye on it as she made the trips between the houses during the periods the water was receding. Between trips, Mrs. Todd counted seven waves. When the eighth arrived it was a towering black mass, higher than the trees on Coconut Island. She ran to the car to make her escape. As she drove away she could hear the wave smashing her house.

Millard Mundy was also one of those who tempted fate but escaped unharmed. In 1946 he, his wife Lucille, and their two children lived in a house on the shore opposite Coconut Island. Employed as a teacher, he was in the bathroom shaving to prepare for school when he saw his neighbor, Mrs. Richardson, letting all her chickens out. Then another neighbor, Rose Chock, shouted that a tidal wave was coming. Mr. Mundy saw that the water in the bay had drained out, leaving the bottom bare. He hurried his pregnant wife and two children into the car in the driveway, but decided that he could not miss this opportunity to take photographs of such a fascinating natural phenomenon. He stood by the water's edge as successive waves came, snapping his photographs. The first few waves did not discourage Mr. Mundy—although Coconut Island was inundated, the nearby shore where he stood did not flood. All at once, he realized that the latest wave had built to an enormous height and presented a terrifying appearance. As it crashed onto the breakwater, boulders the size of automobiles were flung into the air. At that moment, Millard Mundy decided that he and his family were in the wrong place.

Figure 1.11 Tsunami wave washing over Coconut Island in Hilo Bay.

As he accelerated out of the driveway, the house next door was carried out into the sea—colliding with his home as it swept past.

The devastation caused by these waves continued along the coast, wreaking their havoc in Keaukaha.

Keaukaha

At the Hilo Yacht Club, Yoshiko "Charlie" Nakaoka and other staff members were in their quarters behind the club building when the water rose to the level of the steps. Fortunately for them all, Mr. Kennedy, the manager, recognizing the signs of a tidal wave, ushered them into his car, and left at once. As they made their escape through the back roads, the great wave washed away all buildings on the Yacht Club site. At the same time, it crashed into a private home next to the Yacht Club. Unfortunately, the owner, Betty Armitage, had not been as quick to escape as the Yacht Club staff. Waves engulfed her car as she tried to flee. When she tried to leave the vehicle, near the entrance to the Yacht Club, she was swept out to sea by the receding waters.

Just a little farther along the road (where the Mauna Loa Shores condominiums are today), the van Gieson children were getting ready for school. The four van Giesons, eleven-year-old Lenore and ten-year-old Martha, with their brothers E. C. Hebron (eight) and Charles (seven), were in the care of their Uncle Eddie (Laikapu) and Auntie Alice while their parents were in Honolulu. The children went out to the road to wait for the school bus, and Uncle Eddie went with them to catch the newly inaugurated sampan bus service into town. The road was covered in water, so Eddie went down to the shore to investigate. He hurried back, saying "It's a tidal wave," and made them hustle into the house. From the living room (which was on the second story) they looked out through the wooden latticework and saw water foam through the bushes. When it receded, they left the house and ran next door to the Carlsmith home (rented at that time by a young couple), which was on higher ground.

Meanwhile, across the road in a large three-story house on the edge of Lokoaka pond, six-year-old Jeanne Branch and her brother David had not yet got ready for school. They were staying with their grandparents, Charles and Elizabeth Mason, and their Uncle Rod was there too. Their mother had stayed the night in Wainaku (on the other side of Hilo), following an enthusiastic party at the yacht club that had continued in the Baldwin home. When Jeanne and David went out into the street to fetch the newspaper, they saw that the water had flooded the road. Water seemed to be everywhere, but when Jeanne ran to inform her grandmother she was told not to worry, it was just "high seas." Jeanne thought she would be helpful and close the windows (the main living area was on the second floor, and the ground floor consisted of a garage and workshop). When she was closing the kitchen window, she saw that the water was now up to the clothesline. The wave had flooded the garage area and shorted out the car horn, which was blaring loudly. This noise wakened Uncle Rod (who had also been at the yacht club party) and hurt his aching head. He went down into the flooded area and seizing a hammer, attacked the car until the noise stopped. Jeanne and David, still in pajamas, were urged by their

Figure 1.12 Jeanne and David Branch in front of the home of their grandparents, Charles and Elizabeth Mason.

grandmother to get dressed because they had to leave quickly. Jeanne put on a dress and got a paper bag and filled it with her mother's jewelry. Then her Uncle Rod carried her through the water, with David complaining at having to walk as he struggled beside them. As they reached the road they met the van Gieson party, so Rod handed the children over to them. Other people had joined the group, too, and they moved inland together, toward the navigation towers (a series of poles used as navigational aids by aircraft destined for the Naval Air Station or Hilo Airport), the men slashing through the bushes with cane knives. Jeanne wished she had put on more clothes—she had no underwear and no shoes, and it was a slow and painful progress through the sharp lauhala and over the lava rock.

Her uncle had returned to the house, because his father had refused to leave. A retired sea captain and a stubborn man, Charles Mason remained determined to stay with his home, as if it had been his ship. Rod took his mother and the dog next door to the Kennedy house, which was

Figure 1.13 The Mason house, carried away from the front porch and having come to rest in the back yard.

up on a rocky mound. They went up to the top floor and stood on a balcony overlooking the scene. They could see the flooded road and Charles sitting on the lanai (veranda), smoking his pipe. The third wave arrived, swirling around the house from both sides. Just before it struck, Charles jumped through the door into the house, not a second too soon; as the building was lifted from its foundations, the lanai separated from the main structure.

Farther still into Keaukaha, Leslie Waite had been enjoying an early-morning breakfast with his wife and two children. Dr. Waite had just returned from Honolulu, bringing with him some special bakery rolls. The family was sitting at the breakfast table when the maid hurried out of the kitchen. She didn't have the rolls. Instead, she announced that the yardman, Nishiona, was at the door: "He say the ocean look funny." As she spoke, she looked out the window and then screamed, "Tidal wave!" The Waites failed to be alarmed. The maid was only sixteen years old and tended to be excitable. At that point the yardman burst into the room with the news that the narrow strip of road connecting Keaukaha to the main portion of Hilo had been flooded by the ocean. Together, they rushed to pile into the car—the four Waites, the maid, Nishiona, and the one dog they could find. But as he prepared to drive off, Dr. Waite had second thoughts. If the road had flooded already, it might not be safe. Instead, he decided they should walk to the high ground occupied by the navigation towers. When they reached the towers, Mrs. Waite left her baby with the maid and ran back toward the house with the intention of salvaging her spectacles and her purse. As she drew near to her home she heard the roar of the approaching wave, and decided just in time to retreat to safer ground. Meanwhile, her husband had gone to help some elderly neighbors and was across the street from his home when the roaring waters struck it. He watched helplessly as the next wave scattered the wreckage in the street.

Still more residents from the area were gathering by the navigation towers. The Willocks family had been getting out of bed when the first waves arrived. Charles Willocks, manager of the Hilo Iron Works, opened his back door to see 4 feet of water at his back steps and big fish flip-flopping in the grass. He immediately collected his wife, sons, and maid, and ran to the cars—only to find that the vehicles had been jammed together by the wave. So they ran on, up the steep driveway of the house opposite, across the back yard, and inland as fast as they could. They did not stop to watch the waves destroy their house and deposit debris over the driveway they had used for their escape.

The driveway belonged to Lofty Cook. A deep hollow in his front yard became the depository for much of the lumber carried by the wave. The Cook family made their escape in the same direction as the Willocks. From their house (which is raised on a high lava outcrop and, as a result, survived the waves), the Cooks had watched the first waves rolling in before deciding to evacuate.

With them went the Dahlbergs. Their housemaid Virginia had raised the alarm when she discovered that there was no water for her to brush four-year-old Pan's teeth—and no electricity. Pan's father, Bill, went down toward the Willocks' house to see what was happening and was told there were stranded fish on the shore. As he hurried home, the second wave flooded in and he found himself up to his ankles in water. Water was under the house, so they all set off with the Cooks toward the navigation towers: Baby Beamer Dahlberg, Bill Dahlberg with Pan on his shoulders, and Virginia.

Farther along Kalaniana'ole Street, Paul Tallett had been out since before 7 A.M. waiting for the sampan bus. Dressed for work in suit and tie, he waited by his front door. Looking out over the ocean, he noticed that the water was an unusual color, a dark green. He went toward the shore for a better look, and as he did the water receded—tumbling huge boulders in the backwash. He hurried back toward his house, passing Ernie Fernandez, who was raking his yard. "A big sea is coming," he told him. Ernie and his family joined the Talletts at their home, which is raised on an outcrop of rock. From there they saw another wave advance—a mountainous swell that did not break but surged onto the shore. Other neighbors joined them on this high spot of land. Paul thought of the Kais and knew they would need help, as Mrs. Kai was blind. With a rope tied around his waist, Paul tried to reach the Kai home, but the flood was too deep. The sea engulfed the land with a tremendous roar. As the oceanside fish ponds, which had been emptied by the withdrawal of the sea, were refilled by the incoming wave, a tremendous boom filled the air. The watchers saw the

Kwock home demolished. The Kai home was carried out to sea, leveling coconut trees as it went; with the next wave it was lifted and dropped onto the rocks where it burst apart. The people gathered in the Tallett yard were aghast. Would the whole island drown? Before the arrival of the next wave, Paul urged his assembled neighbors (more than twenty people), to evacuate. Along the route Paul discovered young Mrs. Nuhi and her baby stranded in a tree; working quickly, he helped them down and led them to higher ground. Another wave came as they were pushing their way through the trees, but they were far enough from the shore to be safe.

All the ponds in Keaukaha emptied and filled many times that morning. At the Richardson estate, the water passed straight through the house but left it intact. Across the street, Charlie "Chick" Auld had noticed when he got up that morning that the road was flooded. He and his son climbed to the roof of their house, where they watched the progress of the giant waves. One wave rushed in, accompanied by crashes and roars. The next ran up their driveway and took the car from their garage. The wave after that carried their car to Puhele Street, but the house remained undamaged—the Aulds secure on its roof. Mingled with the noise of the waves, they heard the cries of those who had been in the waves' path.

Such cries were echoing up and down the coast. The big waves had all arrived by now. It was nearly 9 A.M. Monday morning, the air still balmy. But the coastline was devastated, laid waste. For many victims of the giant waves, life was changed forever. For others, it was over.

Rescue Efforts

Now the massive task of rescue was about to begin. Some who had been involved in the full fury of the waves would survive, others would not. But heroic efforts were made on their behalf, by many individuals and organizations.

The Hawai'i County Fire Department mobilized very soon after the waves. Robert Napeahi had been on duty at the Central Fire Station, located at the corner of Kinoole and Ponohawai streets. Just after 7 A.M.

people came running up Ponohawai Street, shouting that the Okano Hotel had been flooded by the waves. The firemen went to the roof of the fire station and from there saw the effect of the biggest wave. They could not see the wave itself, but as they watched, the Okano Hotel collapsed.

The firemen were quickly divided into teams. Bob Napeahi took his section to Waiākea. As they made their way from the fire station, they saw destruction everywhere. Buildings on the ocean side of Kamehameha Avenue had been washed across the street. The Cow Palace and the Hilo Theater alone had survived on that side. By Hilo Iron Works, the firemen saw a man stranded at the top of a coconut palm, afraid to attempt the descent. When threatened by an oncoming wave, he had been able to scale the tree, but now he was stuck at the top. People had taken refuge on rooftops, on planks floating in the river, and in skiffs. Many had been caught unaware in their homes and were trapped when the buildings collapsed.

Bob Napeahi helped to rescue one couple through the roof of their home, using his weight to break through the ceiling so that they could climb out. Those rescued were taken by truck to Hoʻolulu race track or to the railroad freight depot, where they were tended by Red Cross workers. But beside the living were many dead: a young woman drowned in her car, a mother and her baby crushed between a large banyan tree and the broken slab of a wall, others floating dead in the river. These bodies were taken by fire truck to a central location to await identification. It was a long and mournful day, with all possible agencies and able individuals involved in the rescue effort.

The police were also trying to bring some order to the chaos. Bob "Steamy" Chow had been backing his car out of his driveway at the corner of Wainaku and Ohai Streets to report for his shift to the police station, when a woman hurrying by told him there was a tidal wave. "Oh yes," he responded, "April Fool!" But when he drove over the arched bridge onto Keawe Street he saw that a span of the railroad bridge had been torn away and had washed up on the river bank behind the Hilo Armory. Then he knew that the "tidal wave" was no joke, but was really happening; he quickly drove to the police station downtown on Kalākaua Street, where his lieutenant instructed him to cover Front Street (Kamehameha Avenue). When he arrived at Kamehameha, he saw a building in the road and debris on the street, with canned goods from the G. Miyamoto store. When he came to the corner of Kamehameha and Wainuenue, he was shocked to discover that the railroad depot had been completely demolished; as he looked toward the lighthouse he noticed a big metal safe that had been lifted onto a wall. He decided he should go inland, and as he drove up Wailuku Drive the second wave struck. People were running all around him so that he could not move his car—the water washed around it and in front of it. The water was not deep enough on the street to cause real problems, but he noticed that the span from the railroad bridge had been lifted, carried farther up the Wailuku River—only just passing under the Keawe Street bridge (where Jim Herkes was watching events)—and deposited on the rock known as Maui's Canoe. Bob knew he had to stop people going down to the bayfront, so he concentrated on doing that for a while: then he went back down to Kamehameha Avenue. At the Amfac building, he saw the canned goods spilled out into the street among the debris and asked the manager if he could eat something, since he had missed his breakfast, then satisfied some of his hunger with some Vienna sausage. It was a short respite in a very long day. For seventeen hours he worked along the ravaged bayfront. He had been asked to go to the Helco cold storage plant because bodies had been taken there. No one knew whether the Wailoa Bridge was safe, so it was after a very roundabout journey that he confronted a shocking scene. In the cold locker he saw the pitiful sight of babies laid out beside old men and women, all victims of the force of the waves.

All who had witnessed the event knew that there was much to be done. Daniel Nathaniel, left swinging from a rafter when the wave receded, dropped safely to the ground. He discovered with joy that his brother had survived also, even though the wave had knocked him down. Five men hastened along the street, looking for their wives. The women had run into the Hilo Meat Market—which had then been flooded—but were able to break out of the back door ahead of the onrushing water. At the nearby Amfac building, the Nathaniel brothers helped the manager to stack up goods knocked down by the waves (and from which Steamy Chow had selected his refreshment). Fortunately, because the waves hit early, there had not been very many people in the shopping area. Peter Alves, who had saved his life by running up Ponohawai Street, returned to the bayfront and helped to pull people from the wreckage of their homes—some living and some dead.

Kapua Heuer, who had watched the terrifying chain of events from her cliff-top home, waited until she was sure that the waves had finished coming in; then she decided that she would be needed at her workplace, the Ordnance Depot at the Naval Air Station. As she went into the street she was confronted by the amazing sight of one of her neighbors, ninety-year-old Mr. Spark, stark naked in the middle of the railroad track. He had been washed out of his bathtub by the rush of water and drowned.

When Kapua arrived at work, her boss asked her to go with a co-worker, Phil Brown, to see what had happened in Keaukaha. There was particular concern about the fate of patients in the tuberculosis hospital at Pu‘umaile.

They drove about a mile and a half along Kalaniana‘ole, as far as Awili Store (now Puhi Bay Store), where the road was blocked. A young boy came riding along on a horse, so they commandeered it. They rode to the home of Kapua's friends, the Codys. At the house Kapua found Sam Cody and asked, "Where's Meredith (Sam's three-year-old daughter)?" He stood stunned and then said, "She's gone." During the big waves he had opened the door and told the maid to run to safety with the little girl. Neither the maid nor the girl were ever seen again.

There was nothing to be done at the Cody house, so Kapua and Phil continued their investigations. At "Four Mile" (James Kealoha Beach Park), the road had been washed out, so she and Phil backtracked around Desha Street, left the horse, and swam across Lokoaka pond. It was not an easy swim—the water was rough and full of debris. They rested for a while by the ruins of the Carlsmiths' house, taking refreshment from a papaya and a bottle of wine they found floating in the water. Then they continued toward Pu‘umaile Hospital, meeting some of the patients, many of whom had been confined to bed for years, straggling along the highway. Some three hundred tuberculosis patients had been marooned in the hospital when the waves had washed out the road and inundated the grounds, and it was feared that subsequent waves might destroy the buildings. With an idea now of what needed to be done, Kapua returned to the Naval Air Station. Lieutenant Commander Barber organized the evacuation of those in Keaukaha, and mobilized the Marines to employ tanks to bridge the water at Four Mile and use rubber dinghies to ferry people across. Soldiers, sailors, and civilians all lent a hand to carry the patients to safety at the navy barracks, where doctors, nurses, and corpsmen provided necessary aid.

Among those ferried across the water at Four Mile were all the people who had taken refuge at the navigation towers. Since they had gathered there, they had been trying to keep up their spirits; most remained calm, with the exception of one elderly woman who was hysterical. Dr. Waite spent his time reassuring the older people, while Mrs. Waite, a teacher at Hilo High School, used the books that the Cook girls had with them to occupy the children.

Bill Dahlberg decided that he had better find out what had happened to his father-in-law, P. C. Beamer, who owned a hardware store on the bayfront. He made his way with some difficulty down to the store to be greeted with the words, “Bill, you’re late.” In the midst of mud and debris, Beamer had opened the store and strung up a sign proclaiming “Open for business.” This proved useful to the Seabees when they needed extra shovels!

During this period, Lofty Cook had returned to his home to fetch food, as well as some milk for the babies. When the family had left early that morning, the table had been set for breakfast, including several portions of cut grapefruit. On his return, he found that the grapefruit had been eaten. He never knew whether it was eaten by a mongoose, a menehune, or a hungry neighbor.

Paul Tallett and Willie Ishibashi had also returned to their homes primarily to look for some of their neighbors who had never arrived at the high ground. They were particularly concerned about the Pua family. As they approached Paul’s home through the floodwater, they found the bodies of old Mrs. Nuhi, her grandson, and two granddaughters. The young Mrs. Nuhi and her baby had been rescued from a tree by Paul and his companion when they made their dash to the navigation tower. Then they heard crying. It was Mrs. Pua; they found her battered and bruised, part of her pajamas torn off. Paul gave her his raincoat and walked with her to the high ground. Once more he returned to his house and collected as much food as he could carry, and, for sentiment’s sake, took an old Hawaiian crook (walking stick) that had belonged to his grandmother.

Lenore and Martha van Gieson saw Paul helping Mrs. Pua as they were walking back toward their home. Like many others, it had been washed from its foundations and jammed against coconut trees near the road. They found out later that although the thick white mugs and plates that they used every day had been cracked by the violence of the waves, the fine Japanese china that had belonged to their grandmother remained intact.

Jeanne and David Branch returned, too, wondering what had happened to their grandparents. Jeanne was astonished and intrigued to see Dr. Waite's house sitting in the road—on top of its furniture! She was delighted to find everyone safe. The Mason house had been lifted from its foundation, but then deposited on the ground. After the waves had stopped, Charles Mason had decided everyone would be hungry and made snacks from Hilo crackers, peanut butter, and jelly, and set them on the wall that separated his property from the Kennedy home.

Everyone had remained safe in the Kennedy house, including a Filipino man who had been washed against a coconut tree by the second wave and would not let go of it when Rod Mason went to fetch him to safety. When persuasion failed, knowing there was little time before the third wave would arrive, Rod knocked him out and took him into the house, earning the lasting gratitude of the man and his family.

The Branch and van Gieson children were excited by the events rather than frightened. They were very willing to go on the rubber boats that had been sent to rescue the residents—although they were not so enthusiastic at the army and navy stations when they had to wait to be reclaimed by their parents. But they were safe, and out of the water.

~

It had been a relatively easy matter to ferry people across the flooded areas of Keaukaha. But in Laupāhoehoe the efforts were sadly hampered by the lack of boats.

David Kailimai, the superintendent of the Hamakua mill, had been at his home in Kūka'iau when the waves struck. About an hour later, one of his workers called him to say that there had been a "big flood" in Laupāhoehoe and trucks were needed to salvage possessions. When he drove down the steep road to Laupāhoehoe and saw the floating debris and the leveled buildings he knew there had been a tidal wave.

He knew also that a boat was essential to save the people—about ten of whom he could see floating in the ocean. No help could be expected from Hilo, since it seemed all too likely to David that their need for boats would be equally great, and that many vessels would have been damaged or destroyed. His own boat was on the other side of the island, in Kona, but he did have the outboard motor at his house. The only boat in the Laupāhoehoe area was a sailboat belonging to a Mr. Walsh. At first Mr. Walsh was not willing for his boat to be used: the sea was still rough and full of debris, the wind was strong, and it seemed likely that damage would occur. Moreover, the boat would have to be cut to accommodate David's motor. It took much persuasion—by the time the boat was launched it was after 2 P.M. The action of the wind and waves had dispersed the people David had seen earlier. He set off to search for survivors accompanied by Libert Fernandez, a young boy named Masaru, and a young soldier, Francis Moku Malani, who was home on compassionate leave. Dr. Fernandez was anxious for the safety of his fiancée, a young teacher. First they found two boys, students from the school; then a seaplane flying overhead directed them to where it had dropped a rubber raft. On the raft they found Marsue McGinnis, the fiancée of Dr. Fernandez. Of the others who had been floating in the water, there was no trace. Marsue McGinnis had been living with three other teachers, Sadie Johnson, Helen Kingseed, and Dorothy Drake, in the first of a group of faculty cottages. The young women had all arrived from the mainland United States the previous September to begin teaching at the school. After the cottage fell apart in the rush of the wave Marsue had a last glimpse of her friends. She saw Sadie and Helen sink almost at once. Neither could swim. Dorothy, her clothes torn off by the wave, was trying to climb onto the roof as it swirled in the water. Then Marsue was dragged down twice as the water "seemed to boil" around her. But she had been able to take a large lungful of air, and remained conscious. When the water calmed she grabbed hold of a log, which floated out to sea. Though numb and dazed, she clung to it tightly, until the rescue plane dropped the raft and she was able to scramble on board. She was the only one of the housemates to survive. In the second cottage, the four student teachers were more fortunate: all of them lived, although one woman, Evelyn (later Mrs. Tommy Crabbe), had a tendon in her leg severed by broken glass from a bookcase.

In the next cottage four more young women survived, including Uniko Yamani. Following a warning shouted from a neighbor, Uniko looked out of the window and saw "an ugly black wave." In her terror, she found herself climbing the room divider, screaming out and calling for her mother. Two of her housemates cowered on the sofa, and the third rushed from the bedroom in response to Uniko's screams—just in time, for the bedroom walls splintered as the wave struck the cottage. In the afternoon Uniko went back to the ruins of her cottage; although the house had been torn in half, the living room remained almost intact. Still in a state of shock, she entered the house. All day she had been worrying about her panty girdle, hanging in the bathroom exposed to view. Uniko found her girdle and took it and her diploma from the ruins. She has no explanation for why she chose those things over the rest of her possessions.

The cottages were found wedged on the wall of a piggery that had probably saved them from complete destruction. It didn't save the pig, however; he had disappeared into the sea.

Herbert Nishimoto remained stranded in the ocean on his improvised raft. After a period of time he sighted two fellow students named Takemoto and Kuhuki and pulled them aboard. At about 1 P.M., the seaplane that had dropped the raft to Marsue McGinnis dropped one to them. They had no fresh water to drink but did recover some coconuts they found floating in the water. They decided to paddle away from land to avoid the large masses of debris. Darkness fell. Between spells of dozing, they tried to locate the coast, but saw only one light. By morning they had drifted more than 10 miles south to Koholā Point. Another plane flew over them, attracting the attention of a young girl on shore who was on the way to her grandmother's house. She saw the raft and ran to tell a group of cane workers who were coming home for lunch. Two strong swimmers went out to tow the raft, whose occupants were too weak to swim to shore themselves. It was a happy ending for Herbert and his companions.

There was a happy ending too for Albino Aguil and his wife Lucy. He was one of the three truck drivers caught in the wave at the dock as they unloaded sugar. One of Mr. Aguil's workmates was swept out to sea and drowned, the other picked up by the S.S. *Brigham Victory.* He himself was injured but not swept out to sea. He was taken to Hilo Hospital for treatment, and in all the confusion no one knew what had happened to him. His wife Lucy was asked to travel to Hilo to see if his body was among those awaiting identification. In the meantime, he had been driven back to Pahala by his supervisor. Lucy experienced some hours of shock and terror until they were reunited, but she then knew the relief of finding him safe.

For many others on the east coast of the island of Hawai'i, that day and the days that followed it were spent counting the cost of the sudden disaster.

The waves from the Aleutians had not only struck the Hawaiian Islands but also surged down the coast of California. At the Coast Guard Station on Half Moon Bay, small boats were floated as far as a quarter of a mile inland. Waves broke over the Half Moon Bay piers and an automobile was hurled into the front of a house. There was even a fatality. At Santa Cruz, two elderly men, Hugh W. Patrick and Cophus Smith, were walking along a cove on the west end of the beach when they were knocked down and engulfed by a wall of water. As the wave began drawing them out to sea, Smith held onto Patrick, but then another wave pulled him free and Patrick was drowned. Smith managed to save himself only by clinging frantically to a rock.

Waves continued across the Pacific, rising as high as 30 feet in the Marquesas Islands. Even 17 hours later they had the power to swamp fishing boats along the coast of Chile, nearly 9,000 miles away.

But it was the Hawaiian Islands that sustained the most damage, and almost all of the fatalities. On the five main islands 159 people were killed by the waves. On the island of Kaua'i the waves swept over 500 yards inland east of Haena. Two women, one clutching her baby, were stranded as rising water surrounded them. Luckily they managed to swim to safety, but many would be less fortunate and seventeen would die on Kaua'i.

On O'ahu, the highest wave reached 36 feet at Makapu'u Point. Just south of there, a man was seen wading through water up to his armpits fleeing his destroyed home and carrying a family picture over his head. The death toll on O'ahu was six.

On Maui homes were destroyed and amphibious tanks just back from the war had floated into a jumbled pile. The highest waves on Maui would reach 33 feet in the gulch at Kahakuloa, but the greatest loss of life would occur just south of Hāna. Of the 14 lives lost on Maui, 10 died in the tiny village of Hāmoa.

But the island of Hawai'i was by far the hardest hit in the entire chain. In Hilo alone, 96 people lost their lives to the sea. At Laupāhoehoe, 25 were killed, including 16 schoolchildren and five of their teachers. One of the children was found tangled in wire among the debris left by the wave. Wrapped in an army blanket, the body was carried to the principal's office, and because she knew all the children Miss Alexandrina was asked to go down to the school and identify it. Her niece Marlene went with her. When the blanket was lifted, Marlene saw the body of her friend Janet De Caires. The force of the wave had stripped off all her clothes, even down to her saddle oxford shoes. And it had taken her brother and sister as well. Of the 159 people killed, only 115 bodies were ever found. The enormity of the loss was brought home to the people of Hilo by the sight of bodies lined up in the streets outside Dodo Mortuary. Kapua Heuer was among the many who went there on the grim mission of searching for a missing friend or relative. She remembers still the horror on the faces of the dead.

There were so many bodies that some had to be moved to the icehouse for storage. They were stacked in such a way that when it was time for burial, the workers found the corpses had frozen together—necessitating the gruesome task of chipping them apart with ice picks. Episodes like this heightened the atmosphere of unreality the disaster had cast over Hawai'i.

Some families experienced both relief and grief. Matilda Moonie's mother had not been in the apartment when the wave struck because she was already at work at the White Star Laundry on Haili Street. Once she realized what had happened she went searching for her family and found Matilda at the Coca-Cola building—although she did not recognize her at first because Matilda had such a bruised and swollen face. She was also to be reunited with another daughter, one son, and her brother. But at the Naval Air Station dispensary she identified the bodies of her mother, her two sons, and her sister. Her baby son's body was not found until one week later, at the breakwater. She identified him by the pants she had sewn for him.

Yet, terrible though each individual tragedy was, the toll in human lives could have been much worse—the timing of the tsunami was good fortune for many. By 7 A.M. most people, although still in their homes, were at least awake. Had the first wave arrived an hour earlier, many people would have been caught asleep in their beds. If the waves had arrived an hour later, when people had left home for the day, downtown Hilo would have been filled with early morning workers and shoppers. In either case the loss of life would have been much greater.

The timing was irrelevant, however, in terms of property damage. Almost 500 homes or businesses were totally destroyed and another thousand severely damaged; the cost of the destruction totaled an estimated $26 million. Structural damage included buildings, roads, railroads, bridges, piers, breakwaters, fishpond walls, and ships.

Frame buildings situated near sea level suffered the most damage: some floated off their foundations nearly intact, while others were demolished where they stood. Many two-story frame buildings suddenly became one-story structures when their ground floors were washed away and the top story left to rest on the foundation.

The densely built bayfront business district in Hilo was almost totally demolished. Nearly every house on the side of the main street facing Hilo Bay was smashed against the buildings on the other side, but those that absorbed the brunt of the waves saved many of Hilo's downtown stores located farther inland.

The railroads in Hilo, as well as along the northern coast of O'ahu, were wrecked. Railroad cars were overturned on O'ahu, Maui, and Hawai'i. The Hilo train station disappeared. The rails of the Hawai'i Consolidated Railway were torn off the rail bed and in some cases were wrapped around trees. In other cases the tracks were moved en masse, probably floated off the roadbed by the buoyancy of the wooden railroad ties.

Jim and Bob Herkes saw an entire span of the steel railroad bridge across the Wailuku River sheared from its supports and washed 750 feet upriver—passing under but not damaging the highway bridge on which they stood. At Kolekole Stream, 11 miles farther north, an entire leg of a high, steel railroad trestle was twisted off its base and carried 500 feet upstream. The Hawai'i Consolidated Railway, already in financial straits, was forced to close down as a result of the extensive damage to its property.

Figure 1.14 Damage to the downtown Hilo business district caused by the 1946 tsunami.

Figure 1.15 Buildings on the ocean side *(left)* absorbed much of the energy, helping to protect those on the opposite side of the street.

Figure 1.16 Locomotive *(top)* surrounded by debris left by the tsunami. The engineer stayed on board sounding his whistle throughout the first wave and then fled to safety. The caboose *(bottom)* washed under a building during the 1946 tsunami.

Figure 1.17 Iron rails shifted by the force of the waves of the 1946 tsunami.

The rights-of-way and bridges on the Hāmākua coast were offered for sale to both the County of Hawai'i and the territory's Highway Division. The offer was turned down by both agencies and the railroad, and all its assets including the rights-of-way, rails, bridges, locomotives, cars, and buildings were sold to a California salvage company for $81,000. The locomotives were cut up, the wooden cars burned, and the scrap iron shipped to the mainland and sold. The steel bridges were to be dismantled, but only two years later, the Territory of Hawai'i turned around and purchased the rights-of-way and remaining bridges for the sum of $310,000, nearly four times what it could have acquired the entire railroad for. The remaining two spans of the Wailuku River railroad bridge were given a new life: they were moved 13 miles down the coast to Kolekole Gulch, where they can be seen today (inverted) as part of the bridge spanning the gulch.

At Hakalau Gulch, 15 miles north of Hilo, the sugar mill suffered damage. Its location at the mouth of the gulch was only about 10 feet above sea level. Coastal highways also were partly destroyed, largely by undercutting. Moreover, as water flooded into the streets of Hilo dozens of automobiles were tossed about and wrecked.

Figure 1.18 Hakalau Valley inundated by the tsunami of April 1, 1946. The Hakalau Sugar Company mill was virtually destroyed by the water, which rose to a height of 20 feet above sea level.

The mile-long breakwater that protects Hilo harbor from normal ocean waves helped reduce the impact of the giant waves, but nearly 60 percent of the structure was destroyed. Giant blocks of stone, some weighing more than 8 tons, were strewn on the bayfront beach like grains of sand. Pieces of coral 5 feet wide were wrenched from the reefs and tossed on the shore to elevations more than 15 feet above sea level.

The commercial piers in Hilo harbor were severely damaged by the force of the waves washing over them and by debris, which acted as waterborne battering rams. Many small boats, including pleasure craft and sampans of the local fishing fleet, were washed ashore and damaged.

The wooden Naniloa Hotel, situated on the peninsula projecting into Hilo Bay, was largely left intact, although it did lose its dining room, boat house, and swimming pier. The hotel was practically the only building along the shoreline not totally destroyed or seriously damaged.

The giant waves from the Aleutians had, in places, swept more than a half mile inland. The withdrawal of the waves between crests exposed the seafloor for a distance of up to 500 feet below the normal shoreline. The water attained heights ranging from an enormous 55 feet in Pololū valley

Figure 1.19 Aerial photograph of the commercial piers at Hilo after the April 1, 1946, tsunami.

on the island of Hawai'i to as little as 2 feet at the village of Milolii on the opposite side of the island. Photographs taken in Hilo show that the tops of breakers were more than 25 feet above the normal water level in the bay as they swept over Coconut Island.

The people of Hilo had already lived through four years of world war fearing possible Japanese attack. The year 1946 should have been free of fear that Hilo might be destroyed. Yet the forces of nature, the catastrophic waves, had brought Hilo to its knees. Not for long, however. The toughness of the Hilo populace soon began to show itself.

The town's seedier element had found unbroken bottles of liquor in among the wreckage. Incredibly, sand was found inside the still sealed bottles. The enormous water pressure from the waves had compressed the corks and injected sand into the bottles. Not to be deterred from par-

Figure 1.20 Residents of Hilo, cheerful and brave, clean up following the disastrous tsunami of 1946.

taking of such a windfall, the town drunks simply strained the contents of the bottles through a sock before consuming the booze.

They were a just a small part of a great army of people searching in the rubble. The entire community valiantly went about the task of cleaning up and rebuilding after the disaster.

The true spirit of friendship shown by so many was much appreciated by those who had been in the path of the waves. The Mundys had been fortunate to escape alive, but when they finally returned to their home three days later they were appalled by the sight of the mud-filled yard and battered house. Fortunately for them, their family doctor, Clyde Phillips, had made a list of all his patients living in waterfront areas, and he visited them all to offer assistance. At the same moment the Mundys arrived at their home, Dr. Phillips had arrived too. He took one look at Lucille's woebegone face, and those of her two small children; then he reached into his pocket and handed over the keys to his Kona cottage. She was able to recover in the sunshine while her husband Millard hosed down their home.

The waves had destroyed so much—but sometimes items of value, sentimental or material, were found in the rubble. The Odachi family, whose home had been washed into the Wailoa and was finally stranded on its bank, were fortunate in being able to walk away. During the cleanup they heard an announcement on the radio that an urn of ashes had been found in someone's basement, still bearing the label identifying the remains as those of June's father. In Keaukaha, the Carlsmith silver was found in Lokoaka Pond next to the Mason house.

Charles Mason, still determined that the ocean should not get the better of him, found a telephone in Hilo on the very day of the disaster and called a contractor in Honolulu to arrange for his house to be restored to its foundations. The van Gieson home was also replaced on its lot, a little farther from the sea, although for some months the children had to travel from Volcano every day to attend school. Many survivors stayed with other family members while their homes were being restored. Wendell Leite and his family lived with relatives in Papaikou while their home in Honoli'i was rebuilt with a 12-foot wall around the property.

Everyone wanted to put their lives back together and be as they were before. But could it all happen again? Many were left troubled by the horrifying events of that day. Some of the students from Laupāhoehoe, like Carol Billena and Yasu Gusukuma, acquired a terror of the sea. Yasu went to stay with her grandma in Mountain View until school reopened, so that she would not be in any danger of hearing the sea—but the nightmares continued nonetheless. Carol should have felt safe in her home high among the cane fields, but in her dreams the sea overwhelmed her home and her family. Neither of them ever lost their fear of the ocean. Months after the tsunami, they went back to the scene of their trauma and attended school there until the high school was moved to its present position in 1952. But they did not want to play at the park. They wondered what many others wondered. How had it happened? Could it happen again?

2

What Is a Tsunami?

As Hilo dug itself out of the rubble, the authorities began the task of determining exactly what had occurred. The media around the world headlined the destruction in Hawai'i. In South America they told of the *"marimoto,"* in France of the *"raz de marée,"* and in Germany of the *"flutwellen."* Newspaper accounts in English told of the "tidal wave"—but we now know that the terrible waves that came from the Aleutians and wreaked havoc in Hawai'i had nothing to do with the tides. Such enormous, destructive waves have been called "seismic sea waves" by scientists, but they are now generally referred to by the term *"tsunami."* Tsunami (pronounced "tsoo-nah-mee") is a Japanese word written with two characters—*tsu,* meaning harbor, and *nami,* meaning wave. Together they are taken to mean "great wave in harbor"—a fitting term, as these giant waves have frequently brought death and destruction to Japanese harbors and coastal villages. For more than two thousand years, the Japanese have recorded the dangers posed by tsunamis, and the awesome power of these waves is depicted in the famous nineteenth-century print by Hokusai.

Tsunamis have, no doubt, visited the shores of the Hawaiian Islands since the islands first formed. In the Hawaiian language, there are two specialized words to describe tsunamis. *"Kai e'e"* is the general term for tsunami waves, and *"kai mimiki"* refers to the withdrawal of the water before a *kai e'e* wave strikes. It is almost certainly *kai e'e* that gave rise to Hawaiian legends about the sea engulfing the land. One such story tells of a love affair between a woman who lived in the sea outside Waiākea, Hilo (an appropriate place for a tsunami legend), and Konikonia, the reigning king of the area. The woman had been lured ashore to sleep with the king, but after four days warned him that her brothers would come looking for her. It seems that her brothers were *pāo'o* fish and in order for

Figure 2.1 Hokusai's *The Great Wave off Kanagawa on the Tokkaido,* from *The Thirty-six Views of Fuji.*

them to search for her the sea would rise. Accordingly, after ten days had passed, "the ocean rose and overwhelmed the land from one end to the other" until it reached the door of Konikonia's house. Many were drowned, but "when the waters had retreated, Konikonia and his people returned to their land."

With the arrival of western missionaries in Hawai'i, destructive tsunamis were chronicled in greater detail, often with religious overtones.

The Tsunami of 1837

A missionary on the island of Maui named Richard Armstrong related what occurred at Kahului on November 7, 1837. At around 7 P.M. on a calm evening, the water began to withdraw from the beach. As the beach widened to more than 120 feet and the reef became exposed, delighted islanders rushed out to pick up stranded fish. A few individuals, who had probably experienced the phenomenon before, concluded that the sea would soon rush in again and fled toward higher ground. But most were caught unawares as a terrifying wall of water rushed back in. One man saw the water coming into his house and, grabbing his child, ran to safety. As he turned his head to look back upon his home, he was astonished to see the "whole village, inhabitants and all, moving toward him, some riding on the tops of their houses, some swimming, all screaming with fright." The tsunami had carried the entire village of 26 grass houses, complete with their inhabitants, canoes, and livestock, some 800 feet inland, dumping them into a small lake.

Fortunately, many were good swimmers and managed to stay afloat. Some even swam to safety with children, the sick, and the aged on their backs. As Reverend Armstrong summed up, "By the blessing of God, all escaped but two at this place. One of these was a mother who was carried out of the flood by her son, safely, as he supposed, rejoicing that he could aid her in such peril. But how was he disappointed when he laid her on dry ground, to find that she had been overpowered by the shock and was dead!"

At Hilo the scene was more terrible than at Wailuku, for about 10,000 people had been assembled at the bay for religious instruction. After spending a long day in church services, the people had either gone home to rest or were gathered along the shore at sunset when the sea began retreating. The English whaling vessel *Admiral Cockburn* was anchored in the bay at the time, and the master stated that a "great part of the bay was left dry." The Hawaiians rushed down in crowds to witness the strange sight, when suddenly a gigantic wave formed and surged toward them.

The Reverend Titus Coan, who witnessed his flock's distress, stated:

> God has recently visited this people in judgement as well as mercy. . . . The sea, by an unseen hand, had all on a sudden risen in a gigantic wave, and this wave, rushing in with the rapidity of a racehorse, had fallen upon the shore, sweeping everything into indiscriminate ruin.
>
> So sudden and unexpected was the catastrophe, that the people along the shore were literally "eating and drinking," and they "knew not, until the flood came and swept them away."
>
> Some were carried out to sea by the receding current. Some sank to rise no more till the noise of the judgement wakes them.

The waves surged into the village at Waiākea, rising to 20 feet above high water. According to the Reverend Coan, the sea crashed upon the shore "as if a heavy mountain had fallen on the beach."

Men, women, and children struggled in the flood, amid their wrecked homes. So violent was the suction as the sea withdrew that even strong swimmers could make little way, and some sank exhausted. But fortunately some were rescued by the boats of the *Admiral Cockburn*. The master, Captain James Lawrence, ordered his sailors to "search for those floating upon the current" and saved them from impending death.

The scene on shore was horrible. About a hundred houses filled with their occupants and guests had been totally demolished and washed away. "Half frantic parents were searching for their children; children weeping for their parents. Husbands running to and fro inquiring for their wives; wives wailing for their departed husbands."

At Hilo four men, two women, and five children had lost their lives, and at a nearby village two women and a child were drowned. But as the good reverend stated: "Had this providence occurred at midnight, when all were asleep, hundreds of lives would undoubtedly have been lost. But in the midst of wrath God remembered mercy."

The lesson taught that day by the Reverend Coan had been: "Be ye also ready." And as he said later, "This event, falling as it did like a bolt of thunder from a clear sky, greatly impressed the people. It was as the voice of God speaking to them out of heaven." It was a lesson the missionaries would not let them forget.

Even the salty Captain Lawrence was impressed by the day's events. According to Coan, "he was a large and powerful man, bronzed by wind and wave and scorching sun, who had thought little of God or the salvation of his soul." But after experiencing the tsunami, "he knelt at the altar and professed to give himself to the Lord." On returning to his ship, he immediately told his officers and crew that he would drink no more, swear no more, and chase whales no more on the Sabbath!

The sea would continue to surge in and out with small waves for a day or more. The tsunami of 1837 would be remembered both as an act of God and as a terrible natural disaster.

The Tsunamis of 1868

The year 1868 was a particularly bad year for tsunamis in Hawai'i. On April 2 a disastrous local earthquake produced a deadly tsunami, and then on August 16 a tsunami from distant Peru struck the islands. Two U.S. Navy ships were involved in this tsunami, and their stories are worth recounting.

One vessel, a sidewheel steam gunboat, the U.S.S. *Wateree*, had been built in 1863 for the Civil War and assigned to the Pacific Fleet. Following the war, the Pacific Fleet was divided into North and South Pacific squadrons. Assigned to the South Pacific Squadron, the *Wateree* had been patrolling the coast of South America and was in port in Arica, near Lima, Peru, on August 16, 1868. Also at Arica that day was the Peruvian gunboat *Americana* and the U.S. Navy storeship *Fredonia*, the latter carrying $2 million worth of supplies to support the South Pacific Squadron.

At 5:05 P.M. a rumbling noise was heard on board the *Wateree,* followed by a trembling motion of the ship. A major earthquake (later estimated to have a Richter magnitude of 8.5)[1] was occurring, and a seaquake[2] had just struck the vessel. Captain James Gillis rushed on deck in time to see the entire city of Arica reduced to a "mass of ruins" in "less than a minute." Captain Gillis then gave orders to secure the gun battery and batten down the hatches. The captain and ship's doctor had just gone ashore to render aid to the earthquake victims when at 5:32 the sea began to rise rapidly. Lieutenant Commander Steyvesant, now in command, released one of the anchors and started the engine. As the vessel swung around in the turbulent water, it took four seamen to control the helm. But the remaining anchor held, and the *Wateree* maintained her position in the harbor. The *Fredonia*, however, could be seen near shore lying on her side with her deck facing the sea. During the turbulent withdrawal of the water, several small boats were drawn past the *Wateree* and their occupants hauled aboard. Then a man was seen drifting past in a mass of bushes and appeared to be drowning. Midshipman Edward Taussig volunteered to render aid and left in the ship's small cutter. No sooner had

[1]The most commonly used scale for measuring the magnitude of earthquakes was devised by C. F. Richter and is known as the "Richter scale." The scale is not linear but logarithmic, so each unit represents a 10-fold increase in ground movement and a 32-fold increase in energy. For example, an earthquake of Richter magnitude 7 would have 10 times the earth shaking strength and release 32 times as much energy as an earthquake of magnitude 6.

[2]The seismic energy transmitted through the water and felt on board ships in much the same way as an earthquake is felt on land.

the cutter left the relative safety of the *Wateree* when a second wave, this one a monster perhaps 90 feet in height, surged into the bay. The officers and men of the *Fredonia* worked bravely and frantically to save their vessel, but she was taken by the wave, carried toward Alacran Island, and smashed to bits on the rocks by the avalanche of water.

The cutter with Midshipman Taussig was carried back by the current toward the *Wateree,* where he was thrown a mooring line. "Houses, railroad cars, men, dogs, trees, and miscellaneous debris" swirled around the vessel. Suddenly, the line holding the cutter broke and Taussig and his coxswain were adrift in the surging current. They were carried in the direction of the Peruvian warship *Americana*. With hope of finding safe refuge on board the warship, both the officer and sailor took oars and rowed. "I never pulled so hard in my life," Taussig would write to his parents a few days later. As they pulled near the Peruvian ship, a line was thrown to them, but at the very moment they grabbed it, the cutter was smashed against the warship and began sinking. Just as the boat filled with water, the two seamen were safely hauled aboard the *Americana*.

Meanwhile, heavy seas were beginning to break over the *Wateree*. Both anchor chains were out to their ends and straining, when suddenly the forward part of the ship, to which the anchor chains were attached, gave way. The *Wateree,* now adrift in the current and headed toward Alacran Island, just missed the rocky island and began drifting rapidly toward shore.

Back on the *Americana*, Midshipman Taussig had stepped into a scene of utter chaos. The captain and about 85 men had just drowned, and the remaining crew, having broken into the liquor stores, were crazy drunk. The lieutenant left in command had tried shouting orders but finally gave up in despair and wept. Waves swept across her deck and the masts began falling overboard. Taussig lashed himself to the shrouds and hoped for the best.

At about 7:20 P.M. both the *Wateree* and the *Americana* were washed ashore and grounded. The *Wateree* had lost only one crewman, the seaman in charge of the captain's gig, who had been on the beach when the largest wave struck and was carried out to sea by the current. But the *Fredonia* had lost all hands, except for two sailors who somehow miraculously survived the destruction of their vessel, and the captain and paymaster, who had gone ashore to render aid in town.

Figure 2.2 U.S.S. *Wateree* shown in the left foreground. The Peruvian cruiser *Americana* is on the right in the background.

The town of Arica had not fared well either. According to Rear Admiral Turner, in command of the South Pacific Squadron, the upper part of the city had "not a single house or wall left standing" and was "a confused mass of ruins," while the lower parts were "perfectly swept clean, even the foundations, as though they had never existed."

The main job of the surviving Americans was now to render aid to the Peruvian survivors and bury the dead. The crew of the *Wateree* went about distributing supplies to the survivors of the disaster. And though the *Fredonia* had been utterly destroyed, many of her stores had fared rather well and were strewn on the beach. The bureau from the paymaster's stateroom, on a lower deck of the ship, had been washed ashore without losing a single drawer. Among the supplies was an ample store of liquor, and it was said that "for three days (after the disaster) even the most humble 'cholo' (Indian) would drink nothing but champagne." Tents were made of whatever material could be scavenged from the beach; several were made entirely from maps of Bolivia.

A careful survey of the *Wateree* proved that she was practically intact, but as she lay 430 yards from the sea it would be impossible to launch her. However, the vessel stayed fully commissioned for several months following the disaster. The crew constructed heads and washing facilities on shore and started a small vegetable garden. Strict naval discipline was maintained even though the ship was high and dry. This, however, did result in some rather unusual naval practices. Instead of using boats to get around, mules became the most useful means of transportation. Mules were used in the local nitrate industry, and following the earthquake, many were to be found roaming the sand dunes. The sailors needed only to "kidnap" and press them into service with the U.S. Navy. To facilitate the use of mules instead of boats, the animals were hitched by lanyards to a lower boom in readiness. When the captain wished to go for a canter among the dunes, the officer of the deck would pass the word to the boatswain's mate, who would call out "First Mule" or "Second Jackass"; the coxswain would then slide down a line to a burro and come alongside to the ship's ladder in readiness for his commanding officer to mount his steed.

Eventually the *Wateree* was sold at auction to a hotel company and used as an inn. Her guns were sold to the Peruvian government. During an outbreak of yellow fever, the ship became a hospital. During the Peruvian-Chilean war in 1880, her hull was bombarded by the Chilean Navy, while her guns were used by the Peruvians to bombard the Chileans. By

Figure 2.3 U.S.S. *Wateree* shown lying among the sand dunes 430 yards from the sea where she was deposited by the tsunami of August 13, 1868. Next to the paddlewheel, note the ship's ladder, used for "boarding mules."

the turn of the century, all that was left were her gaunt ribs rising above the shifting sands.

And what became of Edward Taussig, the courageous midshipman? Those qualities of intelligence and courage that he had displayed so well during the tsunami of 1868 would ultimately lead to his rising through the ranks to become an admiral.[3]

The 1868 earthquake and tsunami resulted in losses of $300 million and a reported death toll of 70,000 in South America. Tsunami waves crossed the Pacific to Hawai'i, where they washed out a bridge in Waiākea, flooded houses on Moloka'i, and caused considerable damage on Maui—but thankfully no loss of life.

The Tsunami of 1877

In 1877 a tsunami disaster would again be visited on the people of Hawai'i. On May 9 a great earthquake occurred off the coast of South America between Peru and Chile. The first waves reached Hilo in the dark before dawn on the morning of the tenth. Sheriff Luther Severance of Hilo recalled the events of the day:

> We have had a great disaster at Hilo. On Thursday morning the 10th at about 4 A.M., the sea was seen to rise and fall in an unusual manner, then at 5 A.M. washed up into nearly all the stores in the front of the town, carrying off a great deal of lumber and all the stone walls makai (on the sea side) of the wharf. . . . But at Waiakea the damage was frightful; every house within a hundred yards of the water was swept away. The steamboat wharf and the storehouse, Spencer's storehouse, the bridge across the stream, and all the dwelling houses were swept away in an instant and now lie a mass of ruins far inland.

The American whaling vessel *Pacific* was anchored in Hilo Bay at the time in 24 feet of water. When the sea receded she was left high and dry,

---

[3]Much of this account was collected by Captain J. K. Taussig, son of Admiral Taussig. In 1922 Capt. Taussig was in command of a cruiser that delivered aid following a tsunami centered 500 miles to the south of Arica. His own experience inspired him to chronicle that of his father.

and when the waters surged back in she was whirled round and round. All expected to see her drag ashore but somehow she survived. Her master, Captain Smithers, like Captain Lawrence forty-four years before, sent his longboats off across the bay to save the helpless people swimming for their lives in the swirling water.

A small hospital on Coconut Island disappeared as the waves washed completely over the island. A small church at Waiākea was floated off its foundation and washed about 200 feet inland; it reportedly "travelled with much dignity, tolling its bell as it went, and was scarcely injured at all, while the principal houses beside it fell in total ruin."

Another account of the events of the morning is found in a letter from the wife of the Reverend Coan. She recalled:

> I was just rousing from quiet slumbers this morning, not long after five, when [someone] knocking at our door hastened me to it. . . . A [giant] wave had swept upon the shore; houses were going down, and people were hurrying "mauka" with what of earthly goods they could carry. . . . Houses were lifted off their under-pinning and some had tumbled in sad confusion and lay prone in the little ponds that remained of the sea in various depressed places. Riders at breakneck speed from Waiakea brought word of still more complete ruin there; the bridge they said was gone!
>
> People were wading in water where their homes had stood half an hour before, gathering up goods soaked by brine.
>
> At Kanae's place, the word was that old Kaipo was missing. Asleep, with Kanae's babe pillowed near her when the wave came upon them, she had wakened, and hastening out of the house found herself in deep water. Holding the little one above her head, she had the courage and strength to keep it safe till the mother swam for it, and then, no one knows how, the old woman was swept out to sea.

Mrs. Coan also relates the story of the remarkable escape of a seventy-seven-year-old English resident of Hilo. "I got caught, sir," he said. "I should have escaped if I hadn't gone back after my money; when I came down-stairs the roller had hit the house, and before I could get out of the door, the house had fallen upon me. I was dreadfully bruised, and you see sir, as the wave took the house inland, it kept surging about with me in it, and getting new knocks all the while." And what of the money—was it saved? "Oh, no, sir, it all went, six hundred dollars. It was all I had, and I am stripped now and I'm past working." Kneeling by the poor man, the Reverend Coan offered an earnest prayer.

Total damage to Hilo was five dead, 17 badly injured, 37 houses destroyed, 163 left homeless and destitute, and 17 horses and mules drowned.

As word of the disaster spread to the neighbor islands, help would come from every quarter. Her Royal Highness Lydia Dominis, the king's sister, came with donations from Honolulu. The tenth of May would long be commemorated as a day of thanksgiving for the aid received.

For many years after the catastrophe a zinc plate could be seen, nailed to a coconut tree a short distance from the beach. The plate, 5 feet above ground level, marked the height to which the waves rose during the 1877 tsunami, more than 16 feet above the sea. The 1877 tsunami would be the most disastrous on record for Hilo until April 1, 1946.

What Is a Tsunami?

But just what is a tsunami and what causes it? In the very simplest terms, a tsunami is a series of waves most commonly caused by violent movement of the sea floor. In some ways a tsunami resembles the ripples that radiate outward from the spot where a stone has been thrown into the water, but a tsunami can occur on an enormous scale.

The movement at the sea floor that causes the tsunami can be produced by three different types of violent geologic activity: earthquakes, landslides, and volcanic eruptions.

Earthquake Tsunamis

Most tsunamis, including almost all of those traveling across entire ocean basins with destructive force, are caused by submarine faulting associated with large earthquakes. The tsunamis that struck the Hawaiian Islands with such force in 1837, 1868, 1877, and 1946 were all associated with large earthquakes.

These are produced when a block of the ocean floor is thrust upward, or suddenly drops, or when an inclined area of seafloor is suddenly thrust sideways. In any event, a huge mass of water is displaced, producing the tsunami. Such fault movements are accompanied by earthquakes, which are sometimes referred to as "tsunamigenic earthquakes." Most tsunamigenic earthquakes take place at the great ocean trenches, where the tectonic plates that make up the earth's surface collide and are forced under each other. When the plates move gradually or in small thrusts, only small earthquakes are produced; however, periodically in certain areas the plates catch. The overall motion of the plates does not stop; only the motion beneath the trench becomes hung up. Such areas where the plates are hung up are known as "seismic gaps" for their lack of earthquakes. The forces in these gaps continue to build until finally they overcome the strength of the rocks holding back the plate motion. The built-up tension (or compression) is released in one large earthquake, instead of many smaller quakes, and these often generate large, deadly tsunamis. In fact, the tsunamis often prove more deadly than the earthquakes themselves. Over the past fifty years, 62 percent of all earthquake-related deaths in the United States have been caused by tsunamis.

Most tsunamis occur in the Pacific Ocean, because the Pacific basin is surrounded by a zone of very active features in the earth's crust: deep ocean trenches, explosive volcanic islands, and dynamic mountain ranges. Frequent earthquakes and volcanic eruptions make the rim of the Pacific basin the most geologically active region on earth. But it is not the only place where tsunamis occur. One of the most devastating tsunamis in history occurred in the Atlantic, produced by the famous Lisbon earthquake of 1755.

The Great Lisbon Earthquake and Tsunami

The enormous earthquake of 1755 occurred on All Saints' Day, November 1, at about 10 A.M. It lasted between 5 and 8 minutes and is estimated to have had a Richter magnitude of at least 8.75. The quake was so strong that its effects were observed as far away as England, where the Viscount Parker, F.R.S., noticed a disturbance of the water in the southwest corner of the moat around Shirburn Castle in Oxfordshire.

Almost immediately after the quake, a powerful tsunami was generated. In Lisbon the sea was seen to withdraw; then a great wave came roaring in, penetrating over half a mile into the city, rushing up streets and inundating houses. Bridges were broken, walls overturned, and great

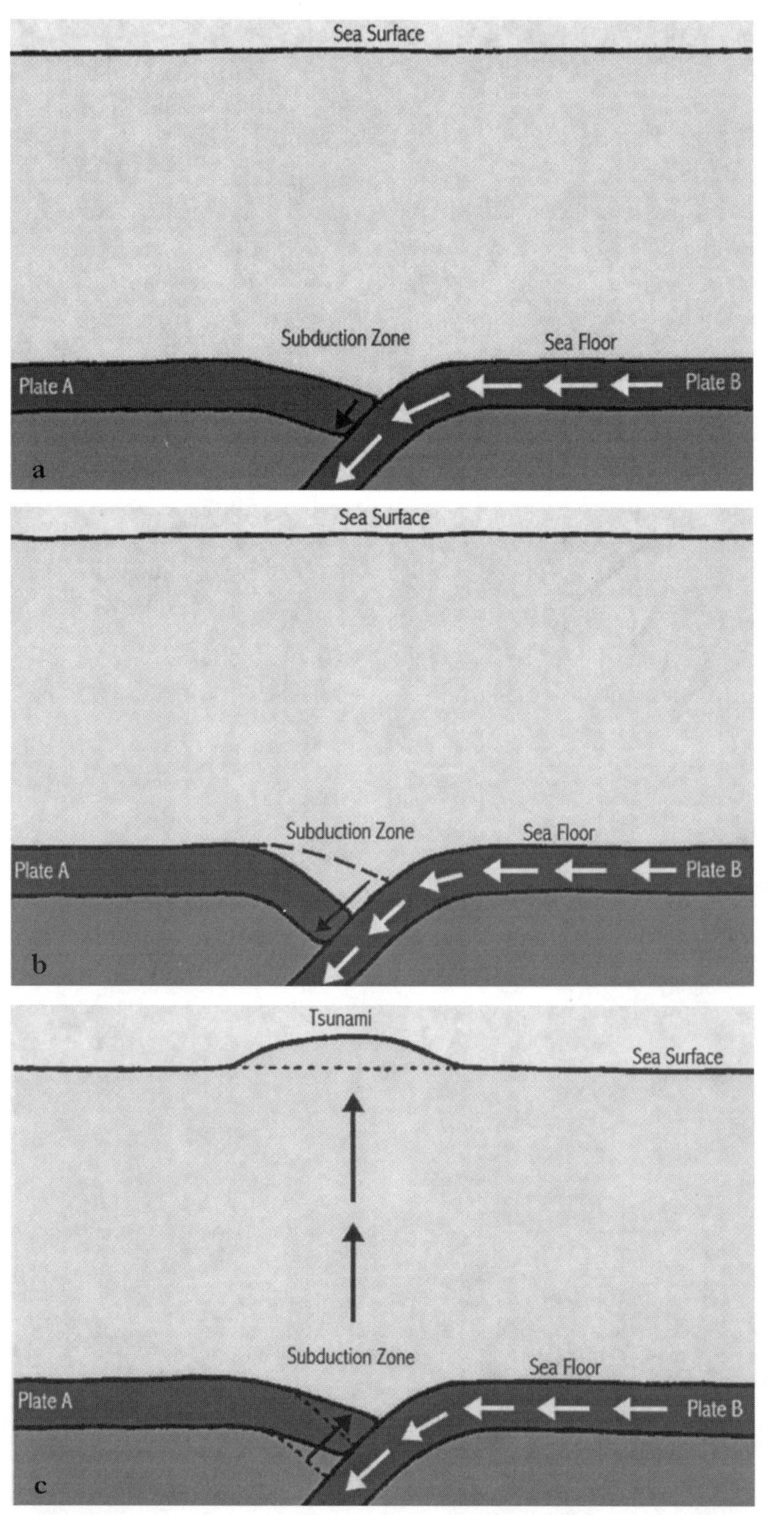

Figure 2.4 Graphic representation of faulting at the ocean floor. Plate A is gradually dragged down into the subduction zone (drawing *a* to drawing *b*) and snaps back, producing an earthquake and generating a tsunami.

piles of debris swept away and carried offshore. At Belem Castle the wave was estimated to have been as much as 50 feet high. One ship's captain said that a wave at least "20 feet high rushed over the quay immediately after the shock." The ships in the river Tagus were said to have been "tossed about as in a violent storm," leaving the harbor a "forest of entangled masts." Many thousands who survived the earthquake were swept away and drowned by the tsunami.

Waves also rushed ashore in Spain and North Africa. At Cadiz, the first wave arrived about an hour after the shock and was said to have been 60 feet high. It tore away portions of the town wall weighing 8 to 10 tons. At Tangier, the first wave is said to have been 50 feet high and to have flowed 1½ miles inland. The tsunami also propagated across the Atlantic. Two hours after the quake, a wave estimated at over 18 feet in height struck Madeira. Similar waves were reported from the Canaries and the Azores. Tsunami waves traveled north to the British Isles, where in Wales a large "head of water" rushed up the river at Swansea, breaking the

Figure 2.5 The Great Lisbon earthquake and tsunami. Artist unknown.

mooring lines of vessels, casting them adrift. Between 2 and 3 in the afternoon a large mass of water poured into the harbor near Cork in Ireland, rising more than 5 feet over the quay. The tsunami even crossed the Atlantic to the Caribbean, where several waves 12 feet high washed ashore in Antigua and flooded streets and wharves in Barbados (nearly 3,600 miles from Lisbon).

But the greatest devastation was in Lisbon itself. The official death toll is usually placed at around 30,000, but the true loss of life was probably closer to 50,000. Upon hearing the news of the disaster, the renowned French philosopher Voltaire was struck with horror by it, and within ten days he had written a long poem about it *(Le Désastre de Lisbonne),* asking how this disaster could be reconciled with a loving God. He argued that its very size was "a terrible argument against Optimism," though he went on to add, "The sole consolation is that the Jesuit Inquisitors of Lisbon will have disappeared with the rest. . . . for while a few confounded rascals are burning a few fanatics the earth is swallowing up both."

Voltaire's attack on optimism set off a literary and philosophical dispute with Jean-Jacques Rousseau, the leading philosopher of the Age of Reason. Rousseau's response was that physical evils were inevitable, and man only added to them by crowding into cities like Lisbon. Voltaire's reply was in the form of his famous work *Candide*. In this ironic masterpiece, Voltaire describes the earthquake and tsunami: "Scarce had they . . . set foot in the city, when they perceived the earth to tremble under their feet, and the sea, swelling and foaming in the harbor, dash in pieces the vessels that were riding at anchor." During the disaster the only truly good man, honest James the Anabaptist, is killed by the tsunami while saving a wicked sailor, who swims ashore unharmed to try and somehow profit from the disaster. Heavy with irony, Voltaire later describes how to mitigate future seismic disasters: " . . . having been decided by the University of Coimbra that burning a few people alive by a slow fire, and with great ceremony, is an infallible secret to prevent earthquakes."

No less ironic than Voltaire's writings were the timing and effects of the earthquake itself. At 10 o'clock on All Saints' Day when the earthquake struck, thousands of people, including most priests, monks, and nuns, were in attendance at morning mass. The earthquake caused most stone buildings to crumble and fall, including 32 churches, 31 monasteries, 15 nunneries, and 60 small chapels. Votive candles lit falling timbers, entombing the faithful in a burning pyre. To add to the irony, among the few buildings safely standing after the disaster were the lightly constructed wooden bordellos of the city. Most of Lisbon's prostitutes, but few of her nuns, survived.

The Virgin Islands Earthquake and Tsunami

Little more than a hundred years would pass before another great earthquake-generated tsunami would strike in the Atlantic—this time in the Caribbean. Three U.S. vessels, in port in the Virgin Islands in mid-November of 1867, were caught up in this tsunami.

But the story has its origins in the Civil War, when Confederate blockade runners and privateers had turned the Caribbean into a lawless sea. Following the war, the United States wanted to put a stop to smuggling and privateering in the Caribbean and therefore wished to have its own naval base in the area. The Virgin Islands, at the time a Danish possession, seemed an ideal location for the base, and the king of Denmark was anxious to sell. An envoy from the king had recently arrived at St. Thomas aboard the Royal Danish mail steamer *La Plata*, which was anchored off Charlotte Amalie. Also anchored in the harbor at Charlotte Amalie were the U.S. Navy side-wheeled steamers *De Soto* and *Susquehanna*. Both vessels had served with distinction during the Civil War.

Off the island of St. Croix was the U.S.S. *Monongahela*. Like the *De Soto* and *Susquehanna*, she was fresh from an illustrious naval career in the Civil War. The ship had been part of Admiral Farragut's squadron that ran past the Confederate batteries at Mobile Bay when the admiral gained fame with his slogan: "Damn the torpedoes. Full steam ahead!" Following the Civil War, she was assigned to the West Indies Squadron, where she came to be anchored in the harbor of St. Croix on November 18.

At 2:45 P.M. on November 18, 1867, a violent earthquake, which would later be estimated to have a magnitude of 7.5, occurred on the seafloor of the Anegada Trough between St. Croix and St. Thomas.

At St. Croix on board the U.S.S. *Monongahela*, Commodore Bissell related, "the first indication we had of the earthquake was a violent trembling of the ship, . . . lasting some 30 seconds, and immediately after the water was observed receding rapidly from the beach; the current changed almost immediately, and bore the ship towards the beach." When the ship was within a few yards of going up on shore, the current suddenly slackened. Commodore Bissell had the sails set and began to make for open water. "A light breeze from the land gave me momentary hope," said the Commodore. Then, suddenly:

> the sea returned in the form of a wall of water 25 or 30 feet high, it carried her over the warehouses into the first street fronting the bay. The reflux of this wave carried her back toward the beach leaving her nearly perpendicular on a coral reef, where she has now keeled over to an angle of 15°. All this was the work of only some three minutes of time.

During the tsunami, an eyewitness on a nearby hill described the *Monongahela* as floating on the top of the great incoming wave on a level with the tops of the houses in Bay Street. The giant waves had been ruinous for the stores along the bayfront. They were inundated and their goods washed out and scattered. The authorities had quickly opened the jail, allowing the prisoners to flee for their lives.

Meanwhile, at St. Thomas, the U.S.S. *De Soto* was anchored with the U.S.S. *Susquehanna* near the entrance to the harbor of Charlotte Amalie; then, according to her captain, David Hall, they "experienced a heavy shock" at 2:50 P.M. Surprisingly, Captain Hall seems to have been prepared for a tsunami. It just so happened that his previous cruise had been with the European Squadron, whose principal port was Lisbon, Portugal. He had become familiar with the great earthquake and tsunami of 1755. Captain Hall felt that the earthquake might be "followed by a tidal wave, so I kept a sharp lookout seaward." He goes on, "I did not have long to wait, for at 3:05 P.M., . . . I saw this immense wall of water." The "sea washed into the harbor about 20 feet high" and the captain "called all hands to save ship." Standing just behind Captain Hall as the tsunami approached was his commander, Commodore Boggs, who was not quite so self assured about the tsunami as the good captain. According to the commodore, the incoming wave had the "appearance of a great bore not less than 23 feet in height." And he continued, "the Island appeared to be sinking."

The captain had the anchors set, but the tsunami proved too much for the chains, which soon snapped. The ship began "drifting all over the harbor," the water "washing in and out with great force." The iron wharf just east of the *De Soto* was destroyed by the tsunami, and the ship was carried over the wharf and impaled on the broken pilings, punching three holes in the starboard side of her hull. Captain Hall wrote in his log, "ship leaking badly. . . . Released all prisoners." The captain finally managed to secure the ship at anchor and listed her to the port side to reduce the leaking.

Not far away, Admiral Palmer was on his flagship, the U.S.S. *Susquehanna*, in his cabin writing when the tremor struck, "accompanied by a sound resembling the grounding of a vessel upon a rough bottom." The admiral calmly resumed writing and "had been seated about ten minutes when the report was brought to me that the sea outside of the harbor had risen and was coming in a huge volume as if to engulf us all." The admiral went on deck in time to witness the tsunami wave approaching. "With a feeling of awe we awaited its arrival; it came rushing on, tumbling over the rocks that formed the entrance, carrying everything before it." The *Susquehanna* managed to ride over "three waves in succession—the anchor chains holding on bravely." But from on board they witnessed small craft in the harbor "lifted up and thrown into the streets. Boats were capsized and men swimming for their lives." As soon as the waters had calmed sufficiently to be safe, the men of the *Susquehanna* rendered aid to those still in danger, her boats picking up and saving several "drowning men."

Anchored near Water Isle, those on board the *La Plata* could see the water "pile up into a wall as it approached the harbor." One of her passengers, Mr. William Maskell, reported that the first wave was a 40-foot-high wall of water as it approached the vessel. The ship rose over each of three waves, but she suffered flooding of the cabins and saloons and breaking of the quarter deck. If the engraving that later appeared in *Harper's Weekly* is accurate, it is a wonder the ship survived at all. The three coaling scows that were fueling the *La Plata* were not so lucky. One was sunk and the other two washed onto the shore over a mile away.

The *Monongahela* lost four men, three of them sailors who had been in her longboats, and the fourth the commodore's own coxswain, who had been in the commodore's gig, which had been crushed under the *Monongahela's* keel. In addition, three or four foolhardy sailors had literally "jumped ship" when the *Monongahela* was washed ashore and had broken arms or legs. The chief boatswain of the *Monongahela* was not on the ship at the time but had his own experience with the tsunami. He had

Figure 2.6 The Danish Royal Mail Steamer *La Plata* anchored off Water Isle, near Charlotte Amalie, St. Thomas, Virgin Islands, during the November 18, 1867, tsunami.

been on shore riding a pony past a local woman and her son, who were gathering wood. When he saw a giant wave approaching, he told the boy to grab hold of the pony's tail and the woman to run for her life. The pony managed to outrun the tsunami, and the boatswain and boy were saved, but the boy's mother was drowned by the wave.

The *La Plata* survived the tsunami, as did the U.S.S. *Susquehanna*. The *De Soto*, though damaged, had also survived and would later rejoin the fleet after repairs in Norfolk.

The *Monongahela* was high and dry, but Admiral Palmer, commander of the squadron, decided to try to refloat her. Several weeks later, however, the admiral died of yellow fever, not living to see the *Monongahela* successfully repaired and refloated six months after the tsunami. She arrived safely in New York on June 1, 1868, and would later serve as the training vessel for the U.S. Naval Academy.

The deal to purchase the Virgin Islands and establish a U.S. naval base was sunk by the disaster. Only three weeks before the earthquake and tsunami struck, the Virgin Islands had been hit by a devastating hurricane, which had damaged nearly every building in the islands. Now with the added destruction from the earthquake and tsunami, the U.S. Senate balked at ratifying the treaty to purchase the islands. Though the Danish government and the islanders themselves were overwhelmingly in favor of the transfer to the United States, the tsunami delayed the ultimate U.S. acquisition of the Virgin Islands until 1917.

The Grand Banks Earthquake and Tsunami

The most recent destructive earthquake-generated tsunami in the Atlantic occurred on November 18, 1929. At 5:02 P.M., Newfoundland standard time, the largest earthquake ever recorded in eastern Canada struck the Grand Banks off Newfoundland. The quake measured 7.2 on the Richter scale and was associated with a massive submarine landslide down the Laurentian Channel into the Cabot Trench. The sediment slide included material from an area measuring 150 by 90 miles. As it moved down slope, it was transformed into a gigantic, fast-moving, underwater mixture of sediment and water known as a "turbidity current." The turbidity current roared down the continental slope, obliterating all in its path. On the sea floor in the area south of Newfoundland lay 12 major transatlantic telegraph cables. In all, 28 separate cable breaks occurred. Some of the breaks happened at the time of the earthquake, while others were produced progressively later as the turbidity current flowed down across the sea floor on its way to the abyssal plains 700 miles away.

Destruction also took place at the sea surface. The Anchor Line ship *Caledonia*, steaming about 120 miles east of Sable Island, Nova Scotia, was suddenly shaken so violently that the ship's master, Captain Collie, feared she either had thrown a propeller blade or was bumping over a sandbank. He ordered the engines stopped and depth soundings taken. Soundings showed the ship to be in over 700 feet of water and no damage could be found to the propeller. Instead, the ship had been struck by a sea quake.

Farther east, three Norwegian steamers were swept by enormous waves. One vessel had her steering gear broken and her lifeboats carried away; another had her smokestack and bridge deck washed overboard; the third lost her rudder. The three ships were near the origin of the tsunami, and it is possible that the waves that damaged the vessels were tsunami waves.

Tsunami waves would also strike the shore. Though tsunami damage was reported from some areas of Nova Scotia, the waves struck with the greatest force principally along a 30-mile stretch of the southern coast of Newfoundland. On the west side of Placentia Bay at the lower end of the Burin Peninsula waves surged over the coast, inundating the shores to an elevation of 100 feet. Unfortunately, the tsunami struck during an unusually high tide and during a raging gale, which magnified the inundation it wrought.

The authorities would report 28 deaths in Newfoundland and one in Nova Scotia. At Lamaline, one man died of injuries and all stores along the waterfront were swept away. At Point au Gaul, the toll was eight lives lost and three houses and 70 other buildings gone. At Taylor's Bay, four were killed and fifteen families left homeless. At Lord's Cove, four lives were lost and all fishing equipment and coal provisions swept away. And at Port au Bras, eight lives were lost, 11 homes destroyed, and 14 small fishing schooners washed out to sea.

Though the tsunami waves from the Grand Banks were recorded as far away as Bermuda and the Azores, the only locations that experienced real damage were in Newfoundland, especially at the heads of converging bays bounded by rocky walls—the sites of most fishing villages. Following the 1929 tsunami, the survivors rebuilt, and in almost exactly the same spots as before.

It would be over twenty years before the Grands Banks earthquake and turbidity current were studied in any detail—and another twenty years before the tsunami would be examined carefully by scientists. Some researchers now believe that the amount of material in the enormous submarine landslide was sufficient to have actually produced the earthquake, and that the slide may have been triggered by the gale or a smaller earthquake. Submarine landslides can indeed generate tsunamis.

Tsunamis Produced by Landslides

Probably the second most common cause of tsunamis is land sliding. A tsunami may be generated by a landslide starting out above sea level and then plunging into the sea, or by a landslide occurring entirely underwater.

Landslides are produced when slopes or deposits of sediment become too steep and the material fails under the pull of gravity. Once unstable conditions are present, slope failure can be caused by storms, earthquakes, rain, or merely continued deposition of material on the slope. Certain environments are particularly susceptible to the production of landslide-generated tsunamis. River deltas and steep underwater slopes above submarine canyons, for instance, are likely sites for landslide-generated tsunamis. The 1929 Grand Banks, Newfoundland, tsunami discussed above was probably generated by a massive submarine landslide down the continental slope.

Fifty years later, a landslide-generated tsunami struck the French Riviera. On October 16, 1979, two waves, each approximately 10 feet high, inundated a 20-mile section of the Mediterranean coast of France stretching from the Italian border to the ancient town of Antibes—an area that includes the coast of Monaco and Nice. The tsunami produced considerable damage and resulted in 10 deaths. To this day, there is some uncertainty about exactly how the tsunami was produced. It is known that a landslide of groundfill material being used to extend an airport runway

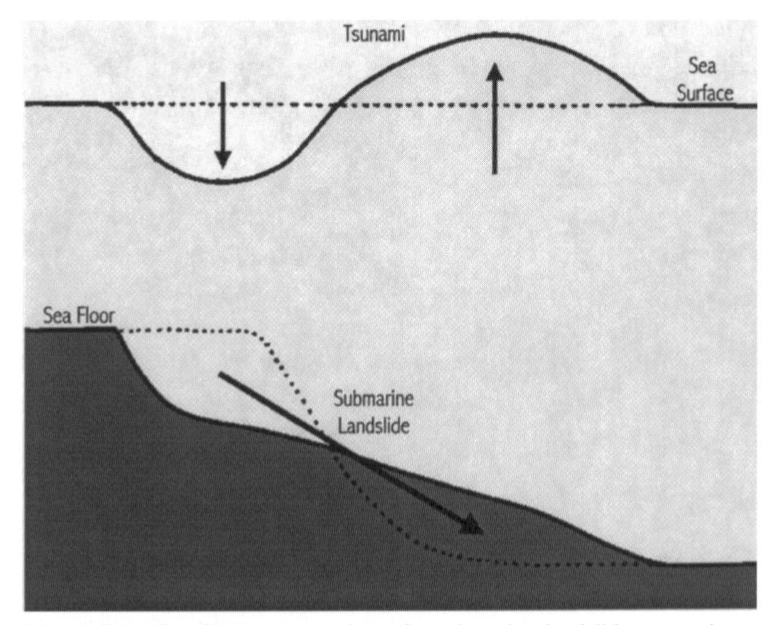

Figure 2.7 Graphic representation of a submarine landslide generating a tsunami. Such landslides may create turbidity currents.

occurred. There were also two large submarine landslides on the continental slope, which continued as turbidity currents, cutting submarine cables at distances of 50 and 60 miles offshore of Nice. The tsunami may have been produced by the airport landslide, or the airport landslide may have triggered the larger, deeper submarine landslides, which could have generated the tsunami.

Another type of area prone to landslide-generated tsunamis are fjords. The oversteepened walls of these glacially carved bays have produced many landslide-generated tsunamis. In Norway, some 210 people have been killed by tsunamis generated by rockfalls and landslides in fjords and glacial lakes. The waves have been as high as 50 feet, with water surging up the slopes to over 200 feet above sea level. Freezing and melting of water in fractures in the rocks triggered an enormous rockfall at Tafjord, Norway, on April 7, 1934. Over a million cubic yards of material fell from a height of nearly 2,400 feet. The tsunami waves were 40 feet high at the opposite end of the fjord. Perhaps the most amazing incidence of a tsunami produced by a landslide in recent times occurred in 1958 in an area of the Gulf of Alaska.

The Lituya Bay Tsunami

On the forbidding stretch of Alaskan coast between Yakutat Bay and Cape Spruce lies Lituya Bay, a T-shaped fjord, with the stem of the T cutting through the coastal lowlands and foothills flanking the Fairweather Range of the St. Elias Mountains. In the center of the stem of the bay is a small, round island, Cenotaph Island, which divides the bay into two channels. The cross piece of the T marks the trend of the active Fairweather Fault and consists of two inlets, which end at the faces of two glaciers, Lituya Glacier to the north and Crillon Glacier to the south. Underwater, the central part of the bay is a U-shaped trench sloping gently downward from the head of the bay to a maximum depth of 720 feet just south of Cenotaph Island. The bottom then rises again toward the outer part of the bay ending at La Chaussee Spit, which nearly closes off the mouth of the bay. The depression that forms Lituya Bay was occupied only recently by a glacier, of which the present Lituya and Crillon glaciers are but remnants. La Chaussee Spit was the end moraine in which this

Figure 2.8 Lituya Bay, Alaska. Cenotaph Island is shown in the center of the bay. The smaller bays of the T are out of view beneath the snow-capped mountains at the top of the photograph. *Key:* r = area of rockslide, d = giant splash, F = maximum distance of tsunami runup, e = small cove where *Edrie* anchored, b = Anchorage Cove where *Badger* and *Sunmore* anchored.

ancient river of ice dumped the silt and rocks carried down from the mountains above.

The weather on the evening of July 9, 1958, was good. There were high scattered clouds in the sky and at this latitude the sun would not set until after 10 P.M. Visibility across the bay was excellent. The bay seemed a perfect refuge from the wild waters of the Gulf of Alaska.

At about 8 P.M., Harold Ulrich and his seven-year-old son Junior entered the bay in their 38-foot fishing boat, the *Edrie*. They anchored for the night in a small cove along the south shore. Orville and Mickey Wagner of Idaho Inlet also entered the bay, but they anchored their 55-foot trawler, the *Sunmore*, in Anchorage Cove on the north shore near the entrance. At about 9 P.M. Bill and Vivian Swanson of Auburn, Washington, entered the bay on their fishing boat, the *Badger*, and anchored near the *Sunmore*.

At 10:16 P.M. the Fairweather Fault began to move. Deep beneath the ice of the glaciers and the water in the inlets, the southwest side of the fault began to sliding toward the north. The earthquake, one of the largest ever to strike North America, would register 8.0 on the Richter scale, and

Lituya Bay was only 13 miles from the epicenter. The fault would be displaced by more than 21 feet laterally and moved up more than 3 feet. The terrible shaking lasted nearly a minute.

Harold Ulrich had just fallen asleep and was suddenly awakened by the violent rocking of his boat. Seismic waves were traveling through the water as a seaquake and shaking his vessel. He immediately went on deck to see what was going on. Bill Swanson was also awakened by the violent shaking of his boat and got up to see what was happening.

A little more than two minutes after the earthquake began, there was a deafening crash. Bill Swanson looked toward the head of the bay, past the north end of Cenotaph Island, and saw what he thought to be the Lituya Glacier, which had "risen in the air and moved forward so it was in sight. . . . It was jumping and shaking. . . . Big cakes of ice were falling off the face of it and down into the water." After a while, "the glacier dropped back out of sight and there was a big wall of water going over the point" (the spur southwest of Gilbert Inlet).

The earthquake had loosened an enormous mass of rock from the northeast wall of Gilbert Inlet. The rock had plunged into the inlet, its impact causing a huge sheet of water to surge over the spur on the opposite side of Gilbert Inlet. Harold Ulrich recalls, "It was not a wave at first. It was like an explosion. . . . The wave came out of the lower part, and looked the smallest part of the whole thing." The explosion was the giant surge, and the wave Ulrich described was a tsunami wave set in motion by the impact of the rockfall. The tsunami wave was now traveling across the bay at a speed of over 100 miles per hour!

The Ulrichs and Swansons looked on frozen in awe at the wave progressing down the bay. About halfway between the head of the bay and Cenotaph Island, the wave extended from shore to shore as a straight wall of water possibly 100 feet high. It was breaking as it came around the north side of the island, but as it approached the *Edrie* on the south side it had a steep, but smooth, even crest, 50 to 75 feet high.

Unable to get the anchor loose, Ulrich let out all 240 feet of chain and frantically started his engine. He put a life jacket on Junior and surrounded him with pillows. As the boat rose with the wave, the anchor chain parted and snapped back around the pilot house. Ulrich grabbed his radio-telephone and yelled, "Mayday! Mayday!—*Edrie* in Lituya Bay—all hell broke loose—I think we've had it—good-bye."

In Anchorage Cove on the north side of the bay, Orville and Mickey Wagner, on the *Sunmore*, had also started their engines and were making a run for the entrance, trying to get outside the bay before the wave overtook them.

Paralyzed by the scene, Bill and Vivian Swanson were still at anchor on board the *Badger*. They saw the wave pass the island about 2½ minutes after it was first sighted. It reached them about a minute and a half later. The *Badger* was lifted up by the wave and carried across La Chaussee Spit. She was riding stern first just below the crest of the wave, like a surfboard, only backward. Bill and Vivian looked down onto the tops of trees growing on the spit: they believe that they were at least a 100 feet above the spit. The wave crest broke just outside the bay, and the boat went down almost vertically with her bow in the air. She hit bottom and foundered offshore. As the *Badger* was going down, Bill and Vivian somehow managed to launch their 8-foot dingy and climb into it. The dingy was nearly swamped and their oars had been washed away, but Bill ripped the thwart loose and began paddling with that. They managed to stay afloat and work their way away from the dangerous waves near the mouth of the bay. The Swansons were rescued an hour and a half later, by Julian Graham aboard his fishing boat, *Lumen*. They were waist deep in icy water and in shock, but still alive.

The *Sunmore* was not as lucky. As the Wagners raced for the pass around La Chaussee Spit, they were caught by the giant wave. The wave lifted and tumbled them over the south side of the entrance, and the boat went down in the ocean just outside the bay.

Meanwhile, Harold and Junior Ulrich, on board the *Edrie*, were carried toward and probably over the south shore, and then, in the backwash, they were propelled back toward the center of the bay. After the first giant wave passed, Harold managed to keep the boat under control in spite of the violent turbulent motion in the bay. For at least 25 minutes, waves up to 20 feet high sloshed back and forth across the bay from shore to shore. Finally, at about 11 P.M. that night, when it seemed possible to survive the passage through the narrow entrance, Harold motored out of Lituya.

The scene at Lituya Bay on the morning of July 10 was unreal. Gilbert and Crillon inlets and the upper part of the main trunk of Lituya Bay were covered with an almost solid sheet of floating blocks of ice. Some chunks were as big as 50 by 100 feet. Across the rest of the bay, huge logs were spread out on the surface and extended out to sea as far as 5 miles from the entrance in a fan-shaped mass of floating timber.

The trees growing on the southwest spur of Gilbert Inlet, directly across from the rockslide, had been cleared to a height of 1,720 feet. Along other parts of the bay, the forest had been felled as far inland as 3,600 feet from shore. The wave had inundated an area of at least 5 square miles, washing away the trees and leaving bare ground. At Harbor Point, a living spruce tree 4 feet in diameter had been cleanly broken off about 3 feet above the ground. Many of the trees felled by the wave were reduced to bare stems, with the limbs, roots, and even bark removed by the turbulent water. A minimum thickness of at least one foot of soil was removed over the entire area, amounting to over 4 million cubic yards of earth carried away by the wave.

The Ulrichs and Swansons had miraculously survived the giant surge and the following tsunami wave, but the Wagners had perished, and no trace of them or the *Sunmore* was ever found.

Yet the loss of life could have been much greater. Twenty-five men near Lituya Bay had come close to joining the list of casualties. A party of eight Canadian mountain climbers had been camped in tents on the shore of Anchorage Cove, at the base of La Chaussee Spit, just before the *Badger* and *Sunmore* came in to anchor for the night. The mountaineers had just returned from the second ascent ever of 15,320-foot Mount Fairweather (British Columbia's highest peak and a boundary with Alaska). Despite a clear evening, their Royal Canadian Air Force pilot was worried about the weather. They boarded their amphibious plane and flew out of Lituya Bay a day early at about 8 P.M. on July 9. This was only a little more than 2 hours before the tsunami wave washed over their campsite.

Also in the area, geologist Virgil Mann of the University of North Carolina and a party of 16 men were camped on the shore of Lake Crillon, 8 miles southeast of Lituya. They were preparing to move to an abandoned cabin on Cenotaph Island the very next day. After the tsunami, not a trace could be found of the old cabin on the slopes of Cenotaph Island, which had been scoured to an elevation of 165 feet.

Even today no one lives at Lituya Bay. Old Russian charts show a large Tlingit village of some 200 inhabitants situated near the entrance to the

Figure 2.9 The large rockslide plunged into the water at the lower right corner of the photo, causing water to surge over the spur opposite *(top center)*, clearing off the forest to an elevation of 1,720 feet. Lituya Glacier is shown to the upper right.

Figure 2.10 Part of the zone of destruction caused by the tsunami wave in Lituya Bay. The width of the zone shown along the right margin of the photograph is 1,700 feet. In other areas trees were felled as far as 3,600 feet from shore. Note that the downed trees near the shore in the foreground with bark removed by the force of the water.

bay, yet by the time a U.S. Coast and Geodetic Survey party entered Lituya Bay in 1874, there were no people living anywhere in the bay. Perhaps they knew the legend of Lituya. Nearly a century earlier, in 1786, the French navigator La Pérouse discovered Lituya Bay and was told a Tlingit legend of a monster who dwelt in the bay. It was said that the monster destroyed all who entered the bay by grasping the surface of the water and shaking it like a sheet. The monster had attacked on July 9, 1958.

Tsunamis Produced by Volcanoes

The violent geologic activity associated with volcanic eruptions can also generate devastating tsunamis. Although volcanic tsunamis are much less frequent than those associated with earthquakes, they are often highly destructive. Villages, islands, and even entire civilizations have disappeared as a result of volcanic eruptions. As with earthquakes, many of the casualties of eruptions are actually victims of the tsunami waves. In fact, of all those listed as killed by volcanic eruptions, nearly a quarter have died as a result of tsunamis. The main reason that volcanic tsunamis can produce such large numbers of casualties is the great distance over which tsunamis travel and cause destruction. For example, tsunamis from the famous eruption of the island volcano Krakatau produced numerous casualties as far as 75 miles away, much farther than could have been directly affected by explosive blast or lava flows.

In historical times, there have been at least 92 major tsunamis produced by volcanic action. About 25 percent of these were produced by earthquakes accompanying eruptions and another 25 percent produced by pyroclastic flows impacting the water. About 20 percent were generated by submarine explosions, about 10 percent by caldera collapse, and the remaining 20 percent by various types of volcanic landslides.

Volcanic earthquakes, unlike those produced by tectonic action, are closely associated with the actual eruption of a volcano. They are usually not very large in magnitude and are only felt locally, close to the volcano. A good example in the twentieth century of destructive tsunamis produced by earthquakes accompanying an eruption is the destruction of the lava dome of Severgin Volcano in the Kuril Islands on January 8, 1933. Three earthquakes were recorded, and each generated a tsunami, the

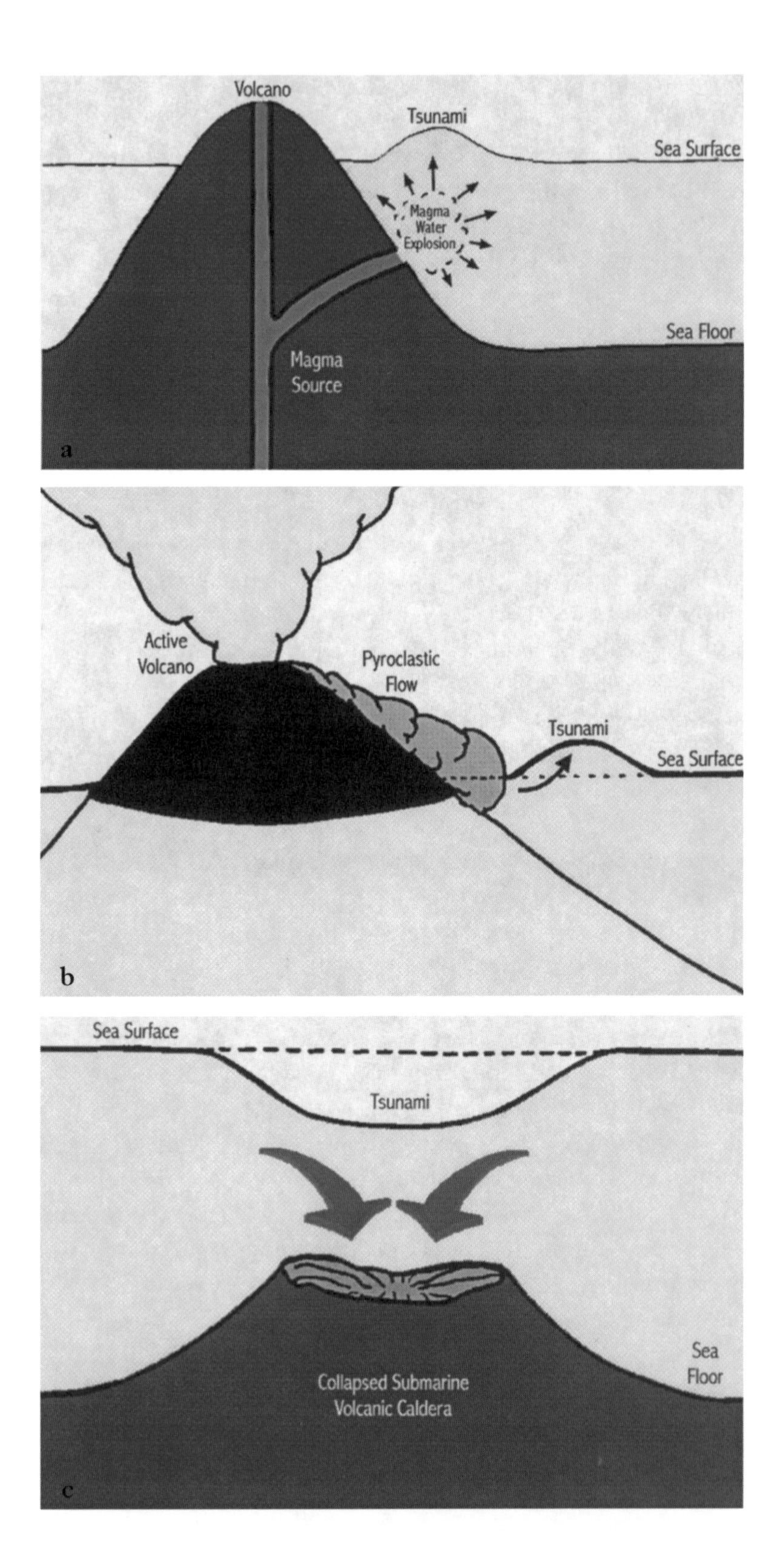

Volcano
Tsunami
Sea Surface
Magma
Water
Explosion
Sea Floor
Magma
Source
a
Active
Volcano
Pyroclastic
Flow
Tsunami
Sea Surface
b
Sea Surface
Tsunami
Sea
Floor
Collapsed Submarine
Volcanic Caldera
c

largest of which was nearly 30 feet high. Probably the largest tsunami in historical times created by a volcanic earthquake occurred on January 10, 1878, in the New Hebrides, where a strong earthquake accompanying the uplift of Yasour Volcano produced a tsunami measuring over 50 feet in height.

Submarine volcanic explosions occur when cool seawater encounters hot volcanic magma (liquid rock). It often reacts violently, producing steam explosions. Underwater eruptions at depths of less than 1,500 feet are capable of disturbing the water all the way to the surface and producing tsunamis. On September 9, 1780, an eruption of Sakurajima Volcano on Kyushu Island, Japan, produced a 20-foot tsunami. On April 11, 1781, Sakurajima again erupted, generating a tsunami that overturned 3 boats, killing 15 people. More recently, a tsunami that was generated by an eruption on Didicas Island in March 1969 killed three people when it struck the northeast coast of Luzon in the Philippines.

Pyroclastic flows have been described by volcanologists as "incandescent, ground-hugging clouds, driven by gravity and fluidized by hot gases." These flows can move rapidly off an island and into the ocean, their impact displacing sea water and producing a tsunami. The famous eruption of Mt. Pelée on May 8, 1902, produced a pyroclastic flow that destroyed the city of St. Pierre in Martinique; the flow's impact with the sea also generated a tsunami. The tsunami was observed as far as 15 miles away at Fort de France. More impressive was the 80-foot tsunami produced by a pyroclastic flow from the eruption of Ruang Volcano in Indonesia on March 5, 1871.

In Alaska the eruption of Mount St. Augustine Volcano on October 6, 1883, produced a pyroclastic flow that created a tsunami with waves nearly 30 feet in height. These struck the town of English Bay, situated about 50 miles east of the volcano, nearly destroying it. An article in an 1884 issue of *Science* magazine stated:

Figure 2.11 Graphic representations of tsunami waves produced by various volcanic processes.
a. Tsunami generated by a submarine volcanic explosion.
b. Tsunami generated by a pyroclastic flow.
c. Tsunami generated by the collapse of a submarine volcanic caldera.
Tsunamis generated by volcanic earthquakes and landslides would be similar to the mechanisms depicted in Figures 2.4 and 2.7, respectively.

> Twenty-five minutes after the great eruption, a great earthquake wave, estimated at from twenty-five to thirty feet high, came upon Port Graham [near English Bay] like a wall of water. It carried off all the fishing boats from the point, and deluged the houses. . . . Fortunately it was low water, or all of the people at the settlement must inevitably have been lost.

In 1898 a geologist with the U.S. Geological Survey recorded a different account of the 1883 event as observed by local natives:

> Traders say here at Katmai that eighteen years ago three natives from Kodiak went with families to St. Augustine Island to spend the winter. . . . the mountain began to shake so violently that they put all their effects in their baidarkas [canoes] and started off. Scarcely were they at the mouth of the bay when an explosion occurred, ashes, boulders, and pumice began pouring down and the barabara [shelters] were buried and the bay filled with debris. At the same time there were many tidal waves, so that the natives nearly perished with fright, yet finally escaped.

Many geologists believe that the tsunamis generated during the great explosive eruption of Krakatau in 1883 were created by pyroclastic flows. The giant, destructive tsunami waves from this eruption will be discussed in detail later in this chapter.

The collapse of a volcanic caldera can generate a tsunami. This may happen when the magma beneath a volcano is withdrawn back deeper into the earth, and the sudden subsidence of the volcanic edifice displaces water and produces the tsunami waves. One of the earliest examples of a tsunami produced by volcanic subsidence dates to the great eruption of Vesuvius in A.D. 79. The Roman writer Pliny the Younger described the events of the morning of August 25: "we beheld the sea sucked back, and as it were repulsed by the convulsive motion of the earth; it is certain at least the shore was considerably enlarged, and now held many sea animals captive on the dry sand." His uncle, Pliny the Elder, admiral of the fleet near Pompeii when Mount Vesuvius erupted, was killed during the eruption while trying to help refugees. A more recent example of a tsunami produced by volcanic subsidence occurred on March 13, 1888, when 2,500-foot-high Ritter Island off New Guinea disappeared into the sea. As the volcano suddenly subsided, it produced a tsunami that washed up 50 feet high on the north coast of New Guinea.

The large masses of rock that accumulate on the sides of volcanoes may suddenly slide down slope into the sea, causing tsunamis. Such landslides may be triggered by earthquakes or simple gravitational collapse. The most catastrophic tsunami produced by volcanic landslides in historical times occurred on the Shimabara Peninsula of Kyushu, Japan, in 1792. Here on the evening of May 21, at about 8 o'clock the townspeople of Shimabara felt two violent shocks, followed by a enormous crash "as if thousands of thunder-bolts were united in a single stroke." A huge portion of nearby Mayeyama (Unzen Volcano) had collapsed and slid into the sea. Although the main mass of material torn from the mountain violently hit the sea, at least part of the foothills had a more gentle slide, as experienced by the watchman on an upland farm located in Uye-no-baru on the flank of the volcano. In the darkness the man felt earth movements, but the farm and its outbuildings seemed to keep their same relative positions. His suspicions were aroused when he began to hear sounds of the sea. When daylight broke, he found that the farm, together with all its buildings, had been carried over a mile down slope toward the ocean.

The violent impact of the landslide into the sea generated tsunami waves at least 30 feet high in Shimabara Bay and there were reports that the water surged up to heights of 180 feet in places along the shore. The coast was devastated over a stretch of nearly 50 miles, and the inundation was made worse by the fact that the tsunami coincided with high spring tides. In all the waves destroyed seventeen coastal villages, killing nearly 15,000 people. It was said that the amount of debris from the town of Shimabara alone was so vast that it formed an offshore embankment over two miles long. In Shimabara town the death toll reached a staggering 5,251. This may have been partly due to a tragic mistake. Shortly after the shocks from the landslide were felt, people were heard screaming on the east side of town. The sentinel on the gate of the castle overlooking the town could see the front of an advancing tsunami wave. As soon as this wave had passed, he quickly closed the castle gates. However, two more tsunami waves followed, both larger than the first. Many of the townspeople were trapped in the streets outside the castle gates and drowned. When Matsudaira Tadahiro, the governor of Shimabara, learned of this, he committed suicide, feeling responsible for the tragic deaths.

More recently, on July 18, 1979, a giant volcanic landslide at Ili Werung

Volcano in Indonesia produced tsunami waves nearly 30 feet high, which washed ashore on Lomblen Island claiming more than 500 lives.

A survey of volcanic tsunamis indicates that the most violent and destructive tsunamis produced by volcanoes are due to pyroclastic flows impacting the sea, or to large volcanic landslides. However, the exact causes of the two most famous tsunami disasters produced by volcanoes are less certain; one of these may have destroyed an entire civilization.

The "Atlantis" Tsunami

A catastrophic volcanic eruption and its ensuing tsunami waves may actually be behind the legend of the lost island civilization of Atlantis. The first written records mentioning Atlantis are in the *Timaeus* by the Greek philosopher Plato. Here he writes of a great island empire destroyed by earthquakes and floods ultimately sinking into the sea. Though Plato is thought to have invented the details surrounding the story of Atlantis to suit his account of an ideal civilization, some scientists speculate that Atlantis was actually the island of Thera and the empire that of the Minoans based on nearby Crete. In 1470 B.C., Santorini, the central volcano in the island group of Thera, underwent a climactic eruption. It is thought that the volcano collapsed, allowing a huge amount of water to enter the magma chamber. When the water came into contact with the hot magma, it was instantly turned into steam, producing a tremendous explosion that also generated prodigious tsunami waves. The explosion may have been as much as 10 times larger than that of the great eruption of Krakatau in 1883, and tsunami waves from 200 to 300 feet high could have been generated. There is evidence that giant destructive waves washed over northern Crete, western Cyprus, and all along the eastern coast of the Mediterranean as far as Tel Aviv and Jaffa. There is also geologic evidence of a wave washing more than 200 miles up the Nile Valley at about this time. Many Mediterranean cultures, including those of the Egyptians, Babylonians, Greeks, and Hebrews, have flood myths that may have been prompted by the Santorini tsunami waves. There is even speculation that it was a tsunami that gave rise to the story of the parting of the Red Sea when Moses and the Israelites were fleeing the Egyptians. The withdrawal of water ahead of a giant tsunami wave would have exposed the sea floor and allowed the Israelites to cross, and then the following crest, coming in as a giant flood wave, would have crushed the pursuing Egyptian army. At any rate, the tsunami generated by the eruption of Santorini certainly ranks as one of the most important to have occurred in the Mediterranean Sea. It appears to have been responsible for the destruction of the Minoan civilization on Crete.

The largest volcanic tsunami in historical times and the most famous historically documented volcanic eruption, and also one of the most thoroughly studied, took place on the other side of the planet in the East Indies. This was the 1883 eruption of Krakatau.

The Destruction of Krakatau

The Indonesian archipelago is one of the most geologically active zones on earth. Major earthquakes originate in the Java Trench and many of the islands were built by explosive volcanoes. The ancient Javanese *Book of Kings* speaks of an eruption with "an enormous blinding flame reaching the heavens" and a mountain exploding into pieces. It goes on to state that "A sea rose and inundated the land. . . . The residents of the Sunda district were drowned and washed away together with their belongings." This definitely sounds like a tsunami—perhaps it was an ancient eruption of Krakatau?

After sleeping for nearly 200 years, Perbuatan, one of the three volcanoes making up the island of Krakatau, suddenly awoke on May 20, 1883, with a series of explosions. The eruption continued throughout the summer, creating vast fields of floating pumice, which reportedly spread far out across the Indian Ocean. Then on August 26 and 27, the eruption came to a spectacular climax as Krakatau blew itself to pieces. The finale was so violent that it has been difficult for geologists to be sure exactly what happened, though a basic chronology has been put together.

The first major tsunamis were generated on the afternoon of August 26. At about 5:30 P.M., waves 3 to 6 feet high struck the town of Anjer, breaking vessels loose from their moorings and damaging the town drawbridge. Then at 7:30 P.M., a tsunami with waves 5 feet high surged into the town of Merak, smashing boats and washing away a Chinese camp. Throughout the night, small explosions from Krakatau continued to be heard and small tsunami waves periodically rolled ashore along the coasts of Java and the Sunda Straits.

The next morning the eruption became even more severe. At about 6:30 A.M., a tsunami wave estimated at 33 feet high crashed ashore at Anjer. The town and harbor were demolished and almost the entire population killed. By about 7:30 A.M., the tsunami had surged across the Sunda Straits into Merek, Ketimbang, and Telok Betong. As the Dutch engineer Abell fled from Merak, he looked back to see a "colossal wave" crash ashore. In the harbor at Telok Betong, the Dutch warship *Berouw* was torn from her anchorage and cast onto the beach.

The eruption continued to intensify and by mid-morning the falling ash was so thick that the entire area was left in nearly total darkness. At 9:58 A.M. a cataclysmic blast occurred. A column of ash soared to a height of 82,000 feet, and the explosion could be heard as far to the north as Manila, as far south as central Australia, and as far away as Sri Lanka and Rodrigues Island, over 3,000 miles to the west. More than two-thirds of Krakatau disappeared in the explosion, including the entire volcanoes of Perbuatan and Danan, and half of Rakata. Most of the material generated during the entire three-month eruption was produced on this single morning. Huge pyroclastic flows poured down into the sea, some creating their own explosions. Thick deposits covered the sea floor, extending for about 10 miles around Krakatau.

It was at this time that the largest tsunami was generated. But there were no witnesses to the event. The entire population of the island closest to Krakatau, Sebesi, a mere 8 miles away, was annihilated. Even farther away, there were few who actually witnessed the largest tsunami come ashore—they had either been killed by earlier tsunami waves or fled into the hills. At 10:18 A.M. the giant tsunami struck Telok Betong, where it refloated the beached vessel *Berouw* and carried her up a narrow valley, leaving the ship some 30 feet above sea level nearly 2 miles inland. All 28 of her crew were killed. At 10:32 A.M. the giant wave swept over the ruins of Anjer and Merak, killing more than 10,000 people. Not a single building was left standing in Anjer. Merak, situated at the head of a funnel-shaped bay, may have experienced the greatest wave anywhere on the coast of Java. Here the tsunami reached an incredible 135 feet above sea level and tossed coral blocks weighing 100 tons up on the shore. At Princes Island off Java and at the lighthouse at Vlakke Hock on the coast

Figure 2.12 Dutch warship *Berouw*, carried into jungle valley 2 miles inland from Telok Betong.

of Sumatra, the water was said to have reached a height of 50 feet. And at Ketimbang, 25 miles across the sea from Krakatau, the largest wave was 80 feet high.

The tsunami waves had swept over nearly 300 coastal towns and villages. The official death toll listed 36,417 killed. But several thousand bodies were swept out to sea and never found. The true death toll will never be known, but probably exceeded 40,000, almost all casualties of tsunami waves.

The tsunami traveled out from Krakatau across the Pacific and Indian oceans. At about 9 P.M. that evening the sea withdrew at Bombay, India. Overjoyed onlookers rushed out on to the exposed tidal flats to pick up stranded fish, narrowly escaping when the sea surged back in. On the northeast coast of Sri Lanka, four adults and three children were washed from a sandbar by a large wave. They were soon rescued by fishermen in boats, but one of the adults, a woman, died two days later of her injuries. She was probably the most distant fatality produced by Krakatau; her death well illustrates how tsunamis have the ability to reach out over vast distances to inflict their destruction.

The tsunami was even recorded in the Atlantic Ocean. It was detected on tide gauges in the Bay of Biscay, and at 9:35 on the evening of

August 28, a half-inch tsunami wave registered on the tide gauge in the port at Le Havre, France, over 10,000 miles away.

The eruption was literally felt round the world, as airborne shock waves repeatedly circled the globe. It became the "catastrophe of the century." A month after the eruption, corpses still littered the beaches of Java and the Sunda Straits. Pumice from Krakatau was reported at sea by ships for two years after the eruption; some of it reached as far as Natal in South Africa, over 5,000 miles away. Over a year later, trees 5 feet in diameter, their roots jammed with pumice, washed ashore on Kosrae in Micronesia, nearly 4,000 miles to the northeast. And almost 4,000 miles to the west, human skulls and bones were discovered a year later, washed up on a beach in Zanzibar.

Just how were these devastating tsunamis generated? Geologists have proposed four different processes that could account for the tsunamis, and all of these took place to one degree or another during the eruption. They are: (1) the collapse of the huge mass of Krakatau Island into the sea; (2) submarine faulting around the margin of the caldera; (3) enormous quantities of volcanic material falling into the sea as pyroclastic flows; and (4) after submergence of the vents, explosions bursting upward to the ocean surface.

Some geologists believe that the explosion theory best fits the evidence. Early speculation that the eruption deposits are laid out in a concentric arrangement was confirmed by a recent underwater survey. This nondirectional distribution has been interpreted to mean that the deposits were laid down by a giant explosion, which could have produced the giant tsunami. Furthermore, at 15 different locations around Krakatau, the first sign of the giant tsunami was a rise in sea level, not a withdrawal. This indicates that the water was displaced upward at its point of origin by an explosion.

Other scientists are not quite so sure and point to an initial drop in sea level observed at some locations on the mainland coast. They point out that sea water rushing in to fill the newly formed submarine cavity created by the subsidence of parts of Krakatau would account for this initial retreat of the sea—hence favoring collapse of the volcano as the mechanism generating the largest tsunami.

Yet another group of scientists favor "several cubic kilometers of pyroclastic flow material entering into the sea immediately after each large explosion" as the mechanism generating the tsunami waves. These flows may have had an internal temperature of up to 1,000° Fahrenheit and could have traveled at a speed of over 300 miles per hour. It has even been suggested that as the flows spread away from Krakatau, the heavier parts of the flow would have sunk under water and continued along the bottom as a submarine flow, while the lighter parts, those less dense than water, would have roared across the surface of the water "cushioned like a hovercraft on their own escaping gases," to quote a Krakatau expert, geologist Steve Self. And a pyroclastic flow can cause a tsunami even if it does not sink. Flows that crossed the Sunda Straits on the sea surface may well have caused a wave merely by pushing water out of their way. And there is evidence that pyroclastic flows from Krakatau actually did flow across the surface of the water, producing an incident know as the "Burning Ashes of Ketimbang."

Across the Sunda Straits from Krakatau was Ketimbang, a village of about 3,000 inhabitants. During the eruption, the Dutch controller and his family, the Beyerinks, had taken refuge in their hillside cabin, when the village was suddenly engulfed by darkness and then fire. Having shut all the doors and windows to the cabin, they saw burning ash begin fountaining up through the cracks in the floorboards. The family was badly burned and one of the children died, along with about a thousand villagers incinerated by the incandescent ash. A pyroclastic flow had evidently crossed 25 miles of ocean from Krakatau.

Not all of the theories are mutually exclusive. In fact, many geologists seem to agree that the most likely explanation for the tsunamis prior to the 10 A.M. giant wave is that they were generated by pyroclastic flows. Many of the small tsunamis during the night were highly localized and could have formed as material ejected from the volcano fell into the sea at various sites around the island. And the giant tsunami, generated at about the time of the culminating blast, could well have been created by the coincidence of rapid pyroclastic flows entering the sea, the sudden slumping of half of the Rakata cone into the actively forming caldera, and the gigantic explosion as cold sea water met molten magma.

Krakatau is one of the most dramatic examples of how one volcanic island can generate numerous destructive tsunamis. The Pacific Ocean is rimmed with coastal volcanoes, including those in Alaska and the Aleutians, Kamchatka, the Kuril Islands, Japan, the Philippines, Indonesia, and

Papua New Guinea. These constitute a very real tsunami hazard. There is also the risk from volcanic tsunamis in the Mediterranean and the volcanic regions of the West Indies.

Some of the largest wave heights ever recorded and highest death tolls have come from tsunamis produced by landslides and volcanoes. In fact, over 58,000 deaths have reportedly occurred as a result of tsunamis generated by landslides and volcanoes. Yet, though tsunamis caused by landslides or volcanic activity may be very large near their sources, and cause great damage there, most such tsunamis decrease in size rapidly, becoming small or even unnoticeable at any great distance. It is usually tsunamis generated by major earthquakes that have enough energy to cross entire ocean basins and inflict destruction at great distances. How do these waves behave at sea far from their points of origin?

Tsunami Waves at Sea

Tsunami waves are very different from other ocean waves. Ordinary waves, which are in fact caused by the wind blowing over the water, affect only the surface of the ocean. Water movement due to these wind-generated waves rarely extends below a depth of 500 feet even in large storms. Tsunamis, on the other hand, involve movement of the water all the way to the sea floor, and as a result their speed is controlled by the depth of the sea. Ordinary wind-generated waves never travel at more than 60 miles per hour and are usually much slower. Tsunami waves, on the other hand, may travel as fast as a jetliner, an astonishing 500 miles per hour or more in the deep waters of an ocean basin. Yet, these same incredibly fast waves may be only a foot or two high in deep water.

Tsunami waves also have much greater wavelengths. Wind waves are rarely longer than 1,000 feet from crest to crest, but tsunami waves are often an incredibly long 100 miles between crests. With a height of 2 or 3 feet spread out over 100 miles, the slope of even the most powerful tsunami would be impossible to see from a ship or airplane. In fact the master of a ship lying offshore near Hilo during the 1946 tsunami stated that he could feel no unusual waves, although he could see great waves breaking on shore.

A popular misconception is that there is only one giant wave in a tsunami. On the contrary, a tsunami may consist of 10 or more waves forming what is called a "tsunami wave train." The individual waves follow one behind the other, anywhere from 5 to 90 minutes apart.

Nearshore

As the tsunami waves move into shallower water and begin to approach shore, they start to change. The shape of the nearshore sea floor, or local bathymetry, has an extreme effect on how the tsunami waves behave.

Tsunami waves tend to be smaller on small, isolated islands, such as Midway or Wake Island, where the bottom drops away quickly into deep water. On large islands, such as the main Hawaiian islands, the tsunami waves have sufficient time to feel bottom and, so to speak, gain a foothold.

As the waves move into the island chain, they begin to be strongly influenced by the bottom and are bent around islands and can even be reflected off the shoreline. The reflected waves may interfere constructively with other waves and create extremely large wave heights in unexpected places.

As the wave forms continue to approach shore, they travel progressively more slowly, but the energy lost from decreasing velocity is transformed into increased wave height. A tsunami wave that was 2 feet high at sea may become a 30-foot giant at the shoreline.

It is commonly believed that the water recedes before the first wave of a tsunami crashes ashore. In fact, the first sign of a tsunami is just as likely to be a rise in the water level. Whether the water first rises or falls depends on what part of the tsunami wave train first reaches shore. A wave crest will cause a rise in the water level and a wave trough causes a recession of the water. Most observers of the 1946 tsunami reported that the first indication of the waves was a withdrawal of the water. However, instruments in both Honolulu and Waimea, Kaua'i, recorded an initial small rise in the water level. The small rise was probably easily overlooked by unprepared observers, whereas the major withdrawal which followed did not escape notice.

Like storm waves, tsunami waves are often more severe on headlands, where the wave energy is concentrated. Unlike ordinary waves, however, tsunamis are also often quite large in bays. In this way tsunami waves, due to their long wavelengths, do resemble tides and may become amplified in long funnel-shaped bays. In fact, some of the greatest wave heights ever observed have occurred in such bays. Unfortunately this has been the case with Hilo Bay, where destructive tsunamis have claimed at least 177 lives.

Seiche

"Seiche" (pronounced "saysh") is another wave phenomenon that may be produced when a tsunami strikes. The water in any basin, be it a bathtub or an ocean basin, will tend to slosh back and forth in a certain period of time determined by the physical size and shape of the basin. This sloshing is known as seiche or *"kai ku piki'o"* in Hawaiian. The greater the length of a body of water, the longer the period of oscillation. The depth of the water also controls the period of oscillations, with greater water depths producing shorter periods. Bathtubs may have a seiche period of 2 to 3 seconds, a swimming pool a period of 8 to 12 seconds, and natural bodies of water from a few minutes to several hours.

A tsunami wave may set off a seiche and if the following tsunami wave arrives with the next natural oscillation of the seiche, water may reach

Figure 2.13 Seiche in Hilo Bay following the 1946 tsunami. Note the small waves crossing Hilo Bay perpendicular to the shoreline.

even greater heights than it would have from the tsunami waves alone. Much of the great height of tsunami waves in bays may be explained by this constructive combination of a seiche wave and a tsunami wave arriving at the same time. Once the water in a bay is set in motion, the resonance may further increase the size of the waves. The impressive wave heights tsunamis produce in Hilo Bay are at least partly related to the seiche effect. Even after the tsunami has passed, the seiche may continue for several days. The dying of the oscillations, or damping, occurs slowly as gravity gradually flattens the surface of the water and as friction turns the back and forth sloshing motion into turbulence.

Bodies of water with steep, rocky sides are often the most seiche-prone, but any bay or harbor that is connected to offshore waters can be perturbed to form seiches, as can shelf waters that are directly exposed to the open sea.

Reefs

The presence of a well-developed fringing or barrier coral reef off a shoreline also appears to have a strong effect on tsunami waves. A reef may serve to absorb a significant amount of the wave energy, reducing the height and intensity of the wave impact on the shoreline itself.

The 1946 tsunami was less severe along the reef-protected northern coast of Oahu than along the unprotected northern coast of the island of Hawai‘i. The most extensively developed coral reef system in the Hawaiian Islands lies inside Kāne‘ohe Bay on the east shore of O‘ahu. In Kāne‘ohe Bay the 1946 tsunami waves were no more than 2 feet high, whereas just outside the bay on the end of the Mōkapu Peninsula, wave heights reached more than 20 feet.

Bores

The popular image of a tsunami wave approaching shore is that of a nearly vertical wall of water, similar to the front of a breaking wave in the surf. Actually, most tsunamis probably do not form such wave fronts; the water surface instead is very close to horizontal, and the surface itself moves up and down. However, under certain circumstances an arriving tsunami wave can develop an abrupt, steep front that will move inland at high speeds. This phenomenon, generally encountered under other circumstances only as a tidal phenomenon, is known as a "bore," or *"kai ko i ka muliwai"* in Hawaiian.

In general the way a bore is created is related to the velocity of shallow water waves. As waves move into progressively shallower water, the wave in front—that is, in shallower water—will be traveling more slowly than the wave behind it. This phenomenon causes the waves to begin "catching up" with each other, decreasing their distance apart (i.e., shrinking the wavelength). If the wavelength decreases, but the height does not, then the waves must become steeper. Furthermore, because the crest of each wave is in deeper water than the adjacent trough, the crest begins to overtake the trough in front and the wave gets steeper yet. Ultimately the crest may begin to break into the trough and a bore is formed.

Bores produced by tides occasionally occur in the mouths of rivers; some of these have been studied intensively. Well-known examples occur on the Solway Firth and the River Severn in Great Britain, on the Petitcodiac River in Maine, near the mouth of the Amazon River in Brazil, and on the lower reaches of the Seine in France. The tidal bore on the Seine is known as the *Mascaret* and was responsible for the death of Leopoldine, daughter of the French writer Victor Hugo. She and her husband, Vacquerie, were drowned near Rouen when the bore knocked them from their small boat into the river. Victor Hugo wrote his tragic poem "A Villequier" as a memorial to the loss of his daughter and son-in-law.

One of the most striking bores occurs on the Qiantang River in China, where it may attain a height of 15 feet. In place of the usual gradual rise of the tide, the onset of high tide is delayed. When it does occur it takes place very quickly, with the rapidly moving wall of water being followed by a less steep, but still quite dramatic, rise in the water level, accompanied by swift upstream currents. In the Qiantang, current speeds in excess of 10 miles per hour are known.

A tsunami can cause a bore to move up a river that does not normally have one. During the 1946 tsunami in Hilo Bay, a bore estimated at 6 to 8 feet high was photographed as it advanced up the Wailuku River. The photograph shows the wave front of the tsunami bore traveling over the

Figure 2.14 Bore on the Qiantang River in China.

Figure 2.15 Bore advancing past the railroad bridge at the mouth of the Wailuku River, Hilo Bay, during the 1946 tsunami.

relatively undisturbed water of the river beneath it. As far as can now be determined, the 1946 bore was of limited extent, restricted to the mouth of the Wailuku River where conditions were favorable for its formation.

Under certain conditions, a much larger and more widespread bore may be formed. A study of the 1960 tsunami found that the third wave to enter Hilo Bay developed a bore that may have reached a height of 35 feet. Bores are particularly common late in the tsunami sequence, when return flow from one wave slows the next incoming wave.

Though some tsunami waves do, indeed, form bores and the impact of a moving wall of water is certainly impressive, more often the waves arrive like a very rapidly rising tide that just keeps coming and coming. In 1946, at some places the water rose gently, flooding over coastal lands with no steep wave front developing. This is how Bunji Fujimoto remembers the waves at Laupāhoehoe. In most areas, however, the advance of the water was accompanied by great turbulence, loud roaring, and

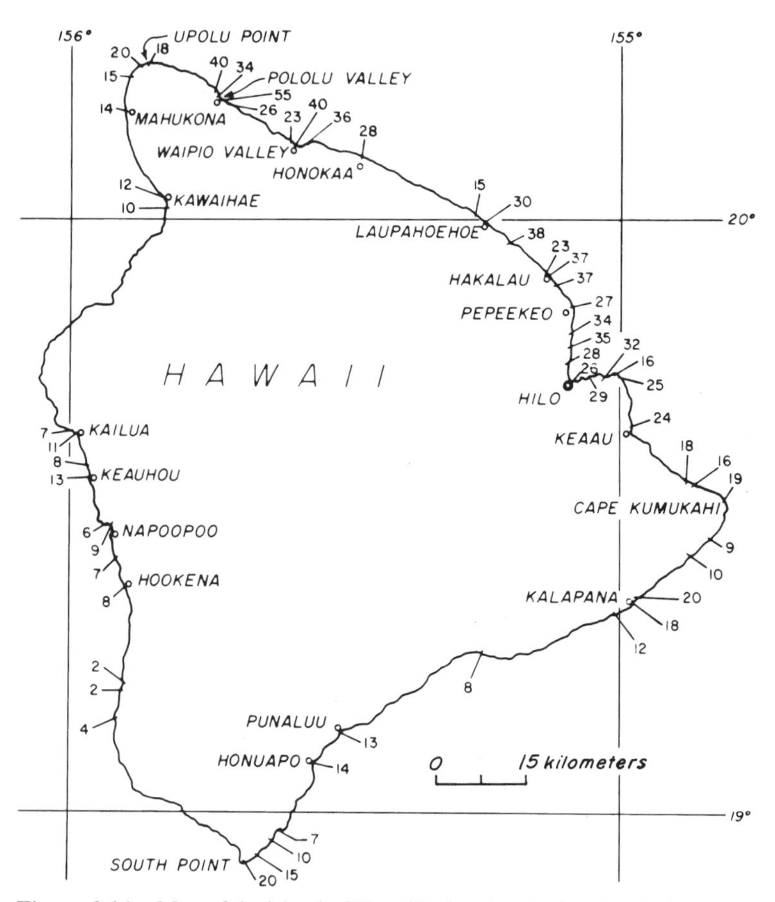

Figure 2.16 Map of the island of Hawai'i, showing the heights (in feet above sea level) reached by the water during the tsunami of April 1, 1946.

hissing noises. The normal wind waves and swells may actually ride on top of the tsunami, causing yet more turbulence and bringing the water level to even greater heights.

Because the height of tsunami waves is strongly influenced by the submarine topography and shape of the shoreline and by reflected waves—and may be further modified by seiches, tides, and wind waves—the actual inundation and flooding produced by a tsunami may vary greatly from place to place over only a short distance. The Aleutian tsunami of 1946 produced waves 30 feet high at Laupāhoehoe, while only a few miles farther up the coast, the waves were 15 feet, half as high.

Just as the wave height and inundation can vary greatly from place to place, so too can the largest wave in the tsunami wave train. In 1946, generally the third or fourth waves were reported to have been the highest and most violent, but at the Waimea River on Kaua'i, the sixth crest was the highest and most destructive.

Damage

Tsunamis like the 1946 one from the Aleutians produce their tremendous destruction in several ways. The sheer force of the moving wall of water in a bore can raze almost everything in its path. It has been estimated that the force of the water in a bore can momentarily attain the enormous pressure of 2,000 pounds per square foot.

Though the image of a bore is the most dramatic, it is the flooding effect of a tsunami that causes the most damage. This was the case in Hawai'i in 1946 and is well illustrated in photographs of the damage to downtown Hilo. Two different terms are often used to describe the extent of tsunami flooding: "inundation" and "run-up." Inundation is the depth of water above the normal level, and is usually measured from sea level at average low tide. Inundation may be measured at any location reached by the tsunami waves. Run-up, on the other hand, is the inundation at the maximum distance inland from the shoreline reached by the tsunami waters.

Even the withdrawal of the tsunami waves can cause significant damage. As the water is rapidly drawn back toward the sea, it may scour out bottom sediments, undermining the foundations of buildings. Entire beaches have been known to disappear as the sand is carried out to sea by

Figure 2.17 Flood damage to downtown Hilo caused by the 1946 tsunami.

the withdrawing tsunami waves. During the 1946 tsunami, the outflow of the water was rapid and turbulent, making a loud hissing, roaring, rattling noise. At several places, houses were carried out to sea with the withdrawing water.

The advance and retreat of tsunami waves causes the water level in ports, harbors, channels, and other navigable waterways to change radically, creating treacherous and unpredictable currents. Boats of all sizes are ripped from their moorings, smashed together, tossed ashore, sunk, or carried out to sea.

Prevention

Can anything be done to prevent the damage and loss of life from destructive tsunamis? After the 1946 tsunami, it was noted that the breakwaters at Kahului on Maui and at Hilo helped reduce the impact of the wave in the harbors behind them. It was suggested that perhaps breakwaters or seawalls could be built as a defense against tsunami waves. In the town of Tarou in Japan, a tsunami wall was built following a devastating tsunami in 1933, which destroyed most of the town and killed one-fourth of the population. The wall was begun in 1934 and now stands 34 feet tall and measures 80 feet thick at its base. But no one knows how effective it will be against tsunamis more than 30 feet high. Though the Hilo and Kahului breakwaters did reduce the waves, most experts agreed that breakwaters or seawalls cannot be built high enough or strong enough to hold the water back completely.

Can waterfront buildings be built to withstand the force of the tsunami waves? After the 1946 tsunami, it was observed that houses elevated on stilts had survived the waves much better than those built directly on the ground. Apparently the water was able to pass under such houses without greatly disturbing them. Reinforced concrete structures (like the Coca-Cola bottling plant) were least affected by the waves. Concrete structural supports were often left standing while weaker nonstructural walls had been carried away.

It is no accident that the hotels that now lie along Banyan Drive in Hilo, facing Hilo Bay, have very open ground floors with high ceilings. Nor is the design meant only to give aesthetic pleasure. The tsunami-resistant design concept for major buildings such as hotels is to have the first living floor area elevated above the potential wave height and to assume that the ground floor and basement will be inundated. The structural walls and columns on the ground floor are designed to resist the impact forces of the waves, while the nonstructural walls between the columns are designed to be expendable as the waves pass through the building.

Following the disastrous tsunami of 1946, much of the downtown bayfront area of Hilo between the Wailuku and Wailoa rivers was left as a parkway. It was hoped that this barrier would serve as a buffer against the destruction of future tsunamis.

But what of the human loss, the tragic deaths caused by the totally unexpected assault of a tsunami? Could anything be done to warn the population of the approach of the waves?

3

The Development of the Warning System

Following the tragic loss of life in Hawai'i as a result of the 1946 Aleutian tsunami, the population wanted to know if anything could be done to warn of the approach of these catastrophic waves. However, the newspaper headlines read "Warning Impossible[,] Geodetic Chief Asserts" and the U.S. Commerce Department denied that its Coast and Geodetic Survey was remiss in not warning the population. Yet, oddly enough, warnings of tsunamis had been issued in Hawai'i during the 1920s and 1930s by the Hawaiian Volcano Observatory (HVO), and the Japanese had their own tsunami warning system in operation by 1941. Why had Hawai'i been caught completely unprepared in 1946, when it had been possible to warn of tsunamis more than 20 years earlier? To answer this question, we need to take a look at the early records of tsunamis in Hawai'i.

During the nineteenth century, numerous tsunamis were reported in newspapers and magazines in Hawai'i, though they were erroneously called "tidal waves." From reading these accounts, it isn't always possible to know whether they refer to a genuine tsunami or to very large wind-generated storm waves. Also, because news traveled slowly in those days, it might take months for news of an earthquake in Alaska or South America to reach Hawai'i. Dates were sometimes mixed up, and consequently the fundamental relationship between earthquakes and tsunamis remained obscure.

Finally, toward the end of the nineteenth century, a seismological station that could record even distant earthquakes was established in Honolulu. It now became possible to associate an earthquake in, for example, South America with a large tsunami in Hawai'i.

In 1912, the Hawaiian Volcano Observatory was established. On the staff were a number of scientists who began to study earthquakes and tsunamis. T. A. Jaggar, the founder of HVO and its director until 1940, began to investigate and report on tsunamis and even researched historical accounts of earlier tsunami events.

Jaggar knew that seismic waves caused by earthquakes are transmitted across the globe in a matter of minutes. A large earthquake in Chile would register on seismographs in Hawai'i hours before a tsunami could reach the islands. Why not use this lead time to warn of an impending tsunami?

In early 1923 Jaggar had the opportunity to witness the earthquake-tsunami relationship firsthand. As he inspected the seismograph at 8 A.M. on the morning of February 3, he noticed the trace of a large earthquake that had been recorded earlier at 5:32 A.M. Hawaiian time. Jaggar quickly calculated that the epicenter would be about 2,500 miles away, possibly under the sea off the Aleutian Islands. The seismic waves had taken only about 7 minutes to travel through the earth to Hawai'i from the Aleutians; if tsunami waves were on the way, they would arrive several hours later. Jaggar notified the harbormaster of the possibility of a tsunami later that day, but his warning was not taken seriously.

The first waves began to arrive at Hale'iwa on the north shore of O'ahu at 12:02 P.M.; there the highest crest would measure over 12 feet. Next the tsunami waves arrived at Maui, where they caused serious damage at Kahului and the withdrawal between crests left a ship grounded on a shoal in the middle of the harbor. At 12:30 P.M. the tsunami struck Hilo, almost exactly 7 hours after the earthquake hit the Aleutians. The largest

Figure 3.1 Matson freighter *Mahukona* aground in Kahului Harbor, Maui, during the February 3, 1923, tsunami.

Figure 3.2 Damage to railroad bridge across the Wailuku River, Hilo, Hawai'i. The bridge collapsed a month after the 1923 tsunami due to undermining of the supports, possibly as a result of the surge.

wave was the third of the series, rising to more than 20 feet at Waiākea. Surging into Hilo Bay, the rising sea carried the local fishing fleet of sampans from their moorage in the Wailoa River into and under the railroad bridge. Most of the fleet of sampans was smashed, and one fisherman was decapitated when his fishing boat was forced under the Wailoa bridge by the power of the tsunami wave. The railway embankment between Hilo and Kūhio was washed away, and the railroad bridge over the Wailuku River was seriously damaged. In addition, wharves and some homes were also damaged.

It was a painful lesson, but it did point out the possibility of protecting life and property by warning of the approach of tsunami waves. Later that same year meteorologist R. H. Finch gave a speech at a scientific meeting in Sydney, Australia. The title was "On the Prediction of Tidal Waves"—a subject that, as Jaggar prophetically stated, "might well be studied to advantage."

Finch pointed out that the time in hours it took tsunami waves to reach Hawai'i was approximately equal to the time in minutes it took the seismic waves to travel here. Using the minutes (for earthquake waves) equals hours (for tsunami waves) rule, he felt that it should be possible to predict the arrival time of tsunami waves from all parts of the Pacific. Finch suggested also that since most seismographs were inspected rather infrequently, some type of "alarm bell" could be attached to the instruments to alert scientists when a large earthquake was registered.

Jaggar felt that the study of tsunamis was now more important than ever. He urged that to accurately predict the time it took the waves to travel from their sources to Hawai'i, it would be necessary to correlate actual earthquakes with the arrival times of tsunami waves. He also thought that by more intensively studying earthquakes and tsunamis, one could determine the earthquake intensity required to generate a tsunami.

Fortunately, most tsunamis are very small and go unnoticed unless registered on a tide gauge. In order to benefit from the study of these small tsunamis, Jaggar traveled to Washington, D.C., to arrange with the U.S. Coast and Geodetic Survey for the establishment of a tide gauge for Hilo.

With tide gauges in Honolulu and Hilo, it was possible to examine the records of changes in water level from these gauges and identify even small tsunami waves of a foot or less. With seismographs, tide gauges, and interested and knowledgeable scientists in Hawai'i, the connection between earthquakes and tsunamis became better known.

In 1933 came another opportunity to test the earthquake-tsunami relationship. At 7:10 A.M. Hawaiian time on March 2, a large earthquake was registered on the seismographs at HVO and on an instrument in Kona operated for the Hawaiian Volcano Research Association by Captain R. V. Woods. Woods was a retired sea captain, who as a volunteer inspected the Kona seismograph daily and sent recordings by mail to HVO weekly. Reading the seismic waves, he calculated the distance to the epicenter as 3,950 miles, possibly off the coast of Japan.

Meanwhile at HVO, seismologist A. E. Jones had come to the same conclusion: a large earthquake had occurred off the coast of Japan. Knowing that a tsunami might possibly have been generated, Jones notified the harbormaster at 10 A.M.; he indicated that waves might begin to arrive about 3:30 P.M. that afternoon.

On the Kona side of the island, Captain Woods notified the Captain Cook Coffee Company at Nāpō'opo'o that waves might come at about 3 P.M. that afternoon. Remembering the 1923 tsunami, workers removed cargo from the dock at Nāpō'opo'o and the Hilo sampan fleet moved out to anchorages in the harbor.

At about noon in Hawai'i, radio news broadcasts announced that a disastrous earthquake had occurred in Japan. In fact, in Japan the Richter magnitude 8.6 earthquake and accompanying tsunami resulted in more than 3,000 deaths, nearly 9,000 homes destroyed, and some 8,000 boats lost. The tsunami waves were highest along the northeast Sanriku coast of the main Japanese island, Honshu. Waves up to 75 feet high were reported on Honshu at Hirota Atsumari. Tsunami waves were also spreading across the Pacific at nearly 500 miles per hour.

On the Kona side of Hawai‘i, facing Japan, the first waves of the tsunami arrived at 3:20 P.M. local time. At first the sea withdrew, exposing wide areas of the sea bottom in the bays at Kailua, Keauhou, Nāpō‘opo‘o, and even at Ka‘alu‘alu near South Point. Canoes and other small craft in Kailua Bay were torn from their moorings and capsized. As the sea returned, walls were washed down, houses were flooded and moved, and in Kailua, a sampan on the marine railway was tossed over the sea wall, landing relatively undamaged on the other side. Of the series of some 10 waves, the last was the most damaging. At Nāpō‘opo‘o the water dropped 8 feet below mean tide and then rose again 9½ feet, for a total vertical range of 17½ feet.

In Hilo the waves began to arrive at 3:36 P.M., but the water rose and fell only a total of 3 feet and caused no property damage.

Thanks to the warnings of Jones in Hilo and Woods in Kona, no lives were lost in Hawai‘i during the 1933 tsunami. Yet these warnings of tsunamis based on earthquakes also led to false alarms. Soon the population began to disregard the warnings, and it was realized that a tsunami warning system based solely on the occurrence of submarine earthquakes was virtually useless. Immediately following the 1946 tsunami, the commander of the Coast and Geodetic Survey stated the case: “Less than one in one hundred earthquakes result[s] in tidal waves and you don’t alert every port in the Pacific each time a quake occurs.”

We now know for a fact that only a small number of undersea earthquakes are accompanied by tsunamis. This is probably because most fault movement either results in lateral, not vertical, motion or takes place deep below the surface of the earth’s crust; thus it causes little or no movement of the sea floor itself, which would displace water. Also, it takes a fairly large earthquake to generate a tsunami. Records of past tsunamis and earthquakes show that it usually requires a quake of at least Richter magnitude 7 to cause a dangerous tsunami in Hawai‘i. It appears that the larger the earthquake the larger the tsunami it can generate, though this generalization is not always strictly true.

In spite of government claims that no warning of the 1946 tsunami had been possible, it was obvious that something had to be done to protect the population of Hawai'i. Both civilian and military sources criticized the Coast and Geodetic Survey for not issuing a warning. After all, as critics pointed out, the seismic waves from the earthquake had been recorded at HVO within minutes after the earthquake struck the Aleutians; consequently, a tsunami could have been predicted.

By coincidence, a number of oceanographers happened to be in Hawai'i in 1946 in connection with the Bikini atomic bomb tests. They observed the tsunami firsthand, and it became the most thoroughly studied tsunami in history. In a scientific paper published in 1947, the well-known researchers G. A. MacDonald, F. P. Shepard, and D. C. Cox stated that they felt that the loss of life from tsunamis could be largely avoided. They recommended that a system of stations be established around the shores of the Pacific and on mid-Pacific islands to observe the arrival of the large, long waves of tsunamis. The arrival of these waves would be reported immediately to a central station, whose duty it would be to correlate the reports and issue warnings to places in the path of the waves. They felt that with such a system it would be possible to give the people of the Hawaiian Islands enough warning of the approach of a tsunami to permit them to reach places of safety. Their ideas combined the advance indication of the "possibility" of the generation of a tsunami, as indicated by the measurement of a large earthquake, with the "confirmation" of a tsunami through measurements from tide gauges lying along the path of the waves. It was felt that such a system could go a long way toward eliminating the problem of false alarms.

For a warning system to be workable, it was necessary to develop a method for quickly and accurately determining the travel time for tsunami waves to arrive in Hawai'i from various earthquake-producing areas around the Pacific. This problem was solved by the preparation of a tsunami travel-time chart for Honolulu in early 1947.

Another obstacle to a workable warning system lay with the seismograph instruments then in use. The visible recording systems for earthquakes in existence in 1946 were of poor accuracy, and the more accurate film recording instruments were read only once daily, when the film was developed. Various new devices were tried out, and in 1947 and 1948 equipment designed by Fred Keller, a scientist living in Pennsylvania, was built and installed at Tucson, Arizona, as well as at College, Alaska, and Honolulu, Hawai'i. These new instruments made highly accurate seismograph records continuously available for inspection, and when a strong earthquake was registered, an alarm was sounded.

Finally, it was essential to establish a rapid, high-priority communications system to transmit reports on the earthquakes and tsunami waves. On August 12, 1948, a tentative plan was approved.

An official tsunami warning system was established by the U.S. Coast and Geodetic Survey and called the Seismic Sea Wave Warning System; the name was later changed to the Tsunami Warning System (TWS). The system was composed of, first, the Coast and Geodetic Survey seismograph observatories at College and Sitka, Alaska; Tucson, Arizona; and Honolulu, Hawai'i; second, tide stations at Attu, Adak, Dutch Harbor, and Sitka, Alaska; Palmyra Island; Midway Island; Johnston Atoll; and Hilo and Honolulu, Hawai'i. The Honolulu (seismic) Observatory, located at 'Ewa Beach on the island of O'ahu, was made the headquarters.

Initially, the warning system was to supply tsunami warning information to the civil authorities of the Hawaiian Islands and to the various military headquarters in Hawai'i for dissemination throughout the Pacific

Figure 3.3 Honolulu Observatory and the Pacific Tsunami Warning Center, headquarters of the Tsunami Warning System of the Pacific, located at 'Ewa Beach, on the island of O'ahu.

to military bases and to the islands in the U.S. Trust Territory of the Pacific. But even as the warning system was being set up, memories of the devastation of 1946 were beginning to fade, at least in the minds of the government agencies back in Washington responsible for funding the system. In fact, during its early years under the Department of Commerce its funding was practically nonexistent. An anecdote recounted by Bernard Zetler, a former manager of the warning system, illustrates the lengths to which the scientists had to go to secure funding in the early 1950s. A colleague of Zetler, Harris B. Stewart, just happened to meet a senator from Hawai'i at a social function and explained the importance of the warning system. The senator offered to write to the secretary of commerce suggesting increased support for tsunami warnings, if Stewart would draft the letter for him. When the secretary received the letter from the senator, he sent it to Stewart and Zetler and asked them to draft a reply. An exchange of several letters followed, with neither the senator nor the secretary realizing that Stewart and Zetler were drafting both ends of the correspondence. In the end the funding was increased slightly, but the warning center continued to be funded on a shoestring compared to most other government agencies.

How the Warning System Works

The system takes full advantage of the tsunami-earthquake relationship. The vast majority of Pacific-wide tsunamis are caused by severe faulting on the ocean floor, and an earthquake of Richter magnitude 7 or greater almost always accompanies the generation of a major tsunami. When a major earthquake does occur, it is recorded by seismographs all over the world within a matter of minutes. Seismograph stations can not only estimate the size of an earthquake but—with reports from 3 or more stations—can also determine the epicenter (the position on the earth's surface directly over the site of origin of an earthquake).

The functioning of the Tsunami Warning System begins with the detection of an earthquake by the network of seismograph stations. Special seismic alarms are set to go off when an earthquake of 6.8 or greater occurs in the Hawaiian Islands or when one greater than 7.0 occurs elsewhere in the Pacific. Scientists at the observatories then rush to their

Figure 3.4 Row of seismographs at the Honolulu Observatory of the Tsunami Warning System. Drum seismographs such as these have now been largely replaced by computer monitors.

instruments and begin interpreting the seismograms. Their readings are almost immediately made available over the Internet to the Honolulu Observatory from the National Earthquake Information Center in Golden, Colorado. Within 15 minutes the Honolulu Observatory has determined the epicenter, but precisely determining the magnitude of the earthquake may take up to an hour.

Tsunami Watch

If the earthquake is strong enough to cause a tsunami and if the epicenter is located close enough to the ocean, a "tsunami watch" is declared. A tsunami watch is automatically issued for all earthquakes greater than Richter magnitude 7.0 occurring in the area of the Aleutians and for all quakes greater than 7.5 occurring elsewhere in the Pacific basin.

Concerned agencies such as State and County Civil Defense, police departments, the American Red Cross, and others are alerted that a tsunami watch is in progress. Local broadcast media may announce the tsunami watch over the airwaves.

Then it is time to confirm if a tsunami has, indeed, been generated, and if so, how big it is. The first positive indication that a tsunami has been generated comes usually from tide-gauging stations nearest the disturbance. Tsunamis appear on the tide gauge records as distinct abnormalities in the normal curve of the rise and fall of the tides.

A Pacific-wide network of communication channels sends out messages requesting those tide stations nearest to the epicenter of the earthquake to monitor their gauges. Trained observers at locations in the path of the tsunami are requested to report to the warning center on wave activity in their areas. A number of unmanned sea level gauges in remote locations also transmit data directly to the warning center.

If the tide stations report negligible waves or no tsunami, then the tsunami watch is canceled. But if the warning center receives reports from tide gauges or observers indicating that a destructive tsunami posing a threat to the population has been generated, then a tsunami warning is issued.

The normal tsunami watch procedure cannot be used in areas immediately adjacent to the point of generation of a tsunami as they are likely to be affected before the confirmation process can be completed. As a result, when an earthquake exceeds a magnitude of 7.5, a tsunami warning is issued immediately for areas within 3 hours travel time from the epicenter of the earthquake, and a tsunami watch is announced for the rest of the Pacific.

Tsunami Warning

When a tsunami warning is declared, the public is informed through the Hawai'i State Emergency Alert System (formerly the Emergency Broadcast System), which broadcasts the warning on all commercial radio stations. State and County Civil Defense agencies implement prearranged plans to evacuate the population from low-lying coastal areas that might be threatened by the tsunami.

Figure 3.5 Computer monitor showing projected advance of tsunami waves across the north Pacific from a source in the Aleutian Islands.

Those areas of the islands that might be subject to the greatest danger have been predesignated, based on studies of the inundation of previous destructive tsunamis. Evacuation routes have likewise been planned by Civil Defense agencies for some areas. Maps of areas in danger of tsunami inundation on each of the Hawaiian islands are printed in the first pages of each island's telephone directory. In the Hilo area evacuation routes are marked by special road signs so as to minimize confusion during an evacuation.

Vessels

As previously noted, vessels offshore may be completely unaffected by tsunami waves if they are in deep, open water. In fact, U.S. Coast Guard vessels always put out to sea upon being advised of a tsunami warning. Once the dangerous waves have passed, they then return inshore to begin rescue operations. Private boat owners are also warned to evacuate their vessels upon the announcement of a tsunami warning. The general rule is to clear the 100-fathom (600-foot) depth line. In many cases, preparing vessels for putting out to sea may need to begin during the tsunami watch

Figure 3.6 Road sign indicating tsunami evacuation route in the Hilo area known as Keaukaha.

stage. Boat owners are further advised that local sea conditions near shore may remain dangerous and unpredictable for several hours after the last major tsunami wave, and it may even take several days to return to normal.

Wave Arrival Time

One of the most important aspects of the entire Tsunami Warning System is its ability to predict the actual arrival time of the first tsunami waves. This capability is based on one of the special characteristics of tsunami waves, their great wavelength. In order to understand just how we determine the arrival time of a tsunami, we need to know a little more about waves in general.

Oceanographers divide waves into various categories based on their wavelength and the depth of water through which they pass.[1] When the water depth is less than one-twentieth the wavelength, the waves are known as "shallow water waves." For example, a wave with a wavelength of 20 feet would be considered a shallow water wave when passing through water less than 1 foot deep. The significance of this is that the speed of a shallow water wave is determined solely by the depth of water. In other words, if we know the water depth, we can calculate the speed of any shallow water wave.

But could tsunamis possibly be considered "shallow water waves" when they cross the deep ocean? Remember that tsunami waves may have a wavelength of more than 100 miles. If the water depth is less than one-twentieth of the 100-mile wavelength, or 5 miles, then the tsunami waves would be shallow water waves. Most of the deep Pacific Ocean basin is less than 3 miles deep, so tsunamis are, indeed, shallow water waves. As a result, in order to determine their speed all we need to know is the depth of the water through which they pass. The depths of most of the ocean basins have been charted; therefore, we can calculate the time for a tsunami wave to travel between any two points in the ocean. In most of the deep sea this time works out to around 450 miles per hour, but it will vary depending on the exact water depth along the path of the tsunami.

A travel-time chart centered on Honolulu is shown in Figure 3.7. Using this chart and knowing the area of origin of a tsunami, we can predict the travel time for the waves to reach the Hawaiian Islands. For example, in the case of a tsunami originating in the Aleutians near Adak, Alaska, the chart shows that it would take 3½ hours for the waves to reach Midway Island and 4½ hours to reach Honolulu. By the time a tsunami watch/warning is declared, and messages are sent out to the emergency management agencies in the warning system, the estimated arrival time of the first wave has already been calculated by computer for all the water level measurement sites reporting to the Warning Center and for many additional

[1]The basic equation for determining the speed of a tsunami wave is $C = \sqrt{gh}$, where C is the wave speed, g is the coefficient of gravity, and h is the depth of water.

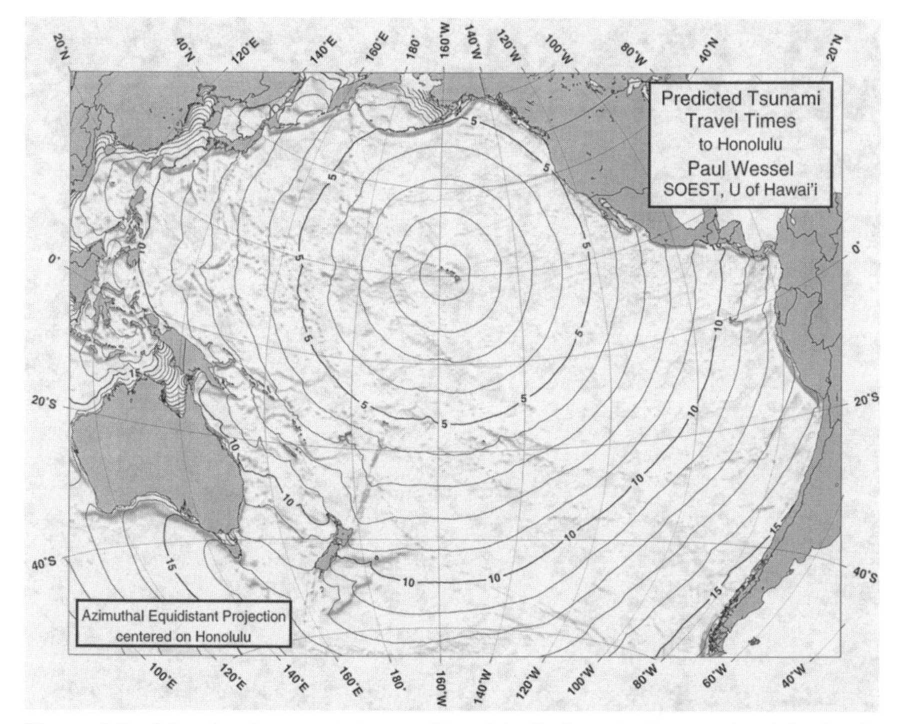

Figure 3.7 Map showing travel times to Honolulu (in hours) of tsunamis originating in the Pacific Ocean.

population centers as well. Evacuation procedures are then coordinated around the predicted arrival time of the tsunami.

The decade of the 1950s dawned with high hopes for the newly established Tsunami Warning System. But there were still a few wrinkles to be ironed out. For example, in Hilo a special device for detecting tsunamis had been mounted on the outer end of the Hilo commercial pier, and when this instrument recorded a tsunami wave, an alarm bell would ring in the Hilo police station. But many people in the islands didn't really trust the new "high technology," so the police chief insisted that someone go down to the regular tide gauge and visually confirm that a tsunami had actually struck. In spite of such minor problems, a network of seismic observatories and tide stations was in place and communication and evacuation procedures had been established. All that was needed to test the system was a Pacific-wide tsunami. The wait would not be long!

4

The Warning System in Action

The 1952 and 1957 Tsunamis

The Tsunami of 1952

Just before five in the morning on November 5, 1952 (1652 Greenwich Mean Time,[1] November 4), a strong submarine earthquake rocked the ocean floor in the far northwestern corner of the Pacific Ocean. Fifteen minutes later, at 1707 GMT, the earthquake alarm at the Honolulu Volcano Observatory in 'Ewa Beach was set off as the shock was registered on seismographs. The warning system immediately went into action. The seismic observatories in California, Arizona, and Alaska all reported in with additional information on the earthquake. In less than an hour, the warning system had determined the earthquake's epicenter to be at 51° north latitude, 158° east longitude, off the southeastern coast of the Kamchatka Peninsula in the Russian Far East. The magnitude registered an impressive 8.2 on the Richter scale.

A tsunami watch was issued. The warning system calculated that the first waves would arrive at Honolulu at 2330 GMT, 1:30 P.M. Hawaiian time. The police and military authorities were notified and constantly

[1]Greenwich Mean Time (GMT) is the time at the prime meridian, i.e. 0° longitude, which passes through the original Royal Observatory in Greenwich, England. Scientists use Greenwich time for events such as earthquakes and tsunamis. Hawaiian time is 10 hours behind Greenwich time, i.e., GMT-10. For example, when it is noon at Greenwich, England, it is only 2 A.M. in Honolulu.

updated by the warning system as reports of the progress of the tsunami waves came in from points closer to its source.

Less than 20 minutes after the earthquake, tsunami waves had struck the coast of Kamchatka, with runup averaging over 20 feet and reaching almost 50 feet above sea level near Cape Pevorotnyi. The settlement of Severokurilsk in the Kuril Islands just to the south of Kamchatka was struck by tsunami waves 65 feet high and totally destroyed. The tsunami was shaping up as a potential monster, and its waves were traveling across the Pacific at a speed of over 500 miles per hour.

At 2005 GMT the tide station at Attu, Alaska, registered a 9-foot drop in the water level. At Dutch Harbor, Unalaska, schools were closed and the population evacuated to higher ground. But the waves at Dutch Harbor came ashore "only" 3 feet high.

In the Hawaiian chain, Midway Island, several hundred miles closer to the origin of the tsunami, was the first to be struck. The first wave surged into the U.S. Navy base at 11:26 A.M. local time. The tide gauge was knocked out of commission by a 6-foot wave, and then a 9-foot wave flooded the island, lifting buildings, washing barges and debris ashore, and depositing sand nearly 1,000 feet up one of the airfield runways.

Just after 1:30 P.M., the tsunami reached Hilo. The tide gauge at the end of Pier 1 recorded a rise in the water level of 4 feet, with no sign of an initial withdrawal prior to the rise. The water continued to rise until it reached a height of approximately 7 feet above "mean lower low water" (MLLW)[2] at 1:49 P.M. Some 16 minutes later, the water level had fallen to about 1 foot below MLLW, the tide gauge showing a total change in water level of some 8 feet. The tide gauge, dampened against very rapid changes in water level, may not have registered the full extent of the inundation. Marks on the sides of the wharf shed on the Hilo commercial pier indicate that the wharf deck was flooded to a depth of nearly 2 feet, some 11½ feet above MLLW.

As the water flooded across the pier, freight awaiting shipment was washed into the bay. A new boat landing for the harbor pilot near Pier 2 was lifted from its foundation by the first wave and totally demolished as it was dropped by the following trough.

---

[2]The reference sea level used in measuring the depths of water on marine charts and tide calendars in most parts of the Pacific Ocean is "mean lower low water" (MLLW). This is the average of the lowest low tides in a particular area.

Figure 4.1 Flooding at Midway Island during the 1952 Kamchatka tsunami.

Moving into Hilo Bay, the tsunami swept over Coconut Island as a 12-foot surge. The walls of buildings on the island were punched out by the force of the water. The small bridge connecting Coconut Island with the shore was floated off its foundation and smashed as it was dropped by the following trough.

At Reeds Bay the water rose as much as 11 feet, destroying a house as the supports were washed out and carrying a fishing sampan 200 feet inland. The mouths of both the Wailuku and Wailoa rivers were flooded to 9 feet above normal, and a small 18-inch bore surged into the Wailoa Estuary during one of the later waves.

The first large waves were followed by a series of ever-diminishing smaller waves with an average period of about 18 minutes. These smaller waves continued for the next several days.

The tsunami continued on across the Pacific striking the west coast of the United States, but most areas would register only tiny waves. Astoria, Oregon, recorded 5 inches and Friday Harbor, Washington, registered a mere 2.4 inches. But Crescent City, California, would record larger tsunami waves than in 1946. Here, the tide gauge recorded 7 feet, and the surge was undoubtably larger away from the protected site of the gauge. Three fishing boats were sunk and heavy concrete mooring anchors were ripped loose and tossed about the harbor.

Even as far away as South America the tsunami was still capable of inflicting damage. In Peru, several houses were inundated by the waves at La Punta, and at Zorritos a 400-foot section of pier was destroyed. At Callao a large number of dead fish washed up on the shore following the tsunami. It is thought that they were asphyxiated as turbulence from the tsunami waves stirred up huge quantities of mud from the bottom, clogging their gills. At Coquimbo in Chile, 500 feet of railroad tracks were flooded and the new customs house was inundated. The tide gauge at Talcahuano, Chile, recorded an impressive 12-foot change in water level due to the tsunami.

On the Big Island of Hawai'i, only the Hilo area received significant damage. Though the waves were too small to measure along many parts of the coast, Hilo Bay would record the largest waves of anywhere in the Hawaiian Islands. Even in nearby Keaukaha the rise in water was so small as to go unnoticed by many. Hilo had proved to be vulnerable not only to Aleutian tsunamis but even to those from the Pacific coast of Russia.

Buildings along Kamehameha Avenue were protected from flooding by the embankment of the new coastal highway, but damage to buildings lying along the shoreline between the Wailoa River and the breakwater was extensive. Several fishing sampans were beached, and others in a nearby boat yard were damaged.

The waves of the 1952 tsunami were not nearly as high as in 1946 and according to observers, instead of "rolling in with a steep, forceful impact, the waves were mostly just a gentle rise of the water." Lawns were flooded, telephone lines knocked down, automobiles marooned, and in Honolulu Harbor a cement barge was hurled against a freighter. But the death toll lay at six cows! Not a single human life was lost, and property damage was estimated at less than $800,000.

The 1952 Kamchatka tsunami would be one of the most thoroughly studied to date and was recorded on more tide gauges than any prior tsunami in history. It was also the most severe tsunami since the establishment of the warning system and proved that the warning system worked. There can be no doubt that the warnings saved lives. However, there were a couple of disturbing events during the 1952 tsunami. In Honolulu,

Figure 4.2 Boat knocked off blocks in Hilo boatyard during 1952 tsunami.

sightseers "stampeded toward the beach instead of away from it," displaying their ignorance of the potential danger of the waves. In addition, sources not connected with the warning system issued statements that the first or second wave crests were the dangerous ones, and that after those the danger was over and people could safely return to their homes or places of business in threatened coastal areas. In fact, the warning system received a report from the commander of the ship *Hawaiian Sea Frontier* warning that the fourth wave had done the most damage at Midway Island, and suggested caution against "securing" early. And as you will recall, studies of the 1946 tsunami showed that at different localities the largest waves ranged from the third to the eighth. The only safe thing for residents of coastal areas to do is to stay away from the shore until at least two hours after the waves stop coming in.

The Tsunami of 1957

More than ten years had passed since the disastrous Aleutian tsunami of 1946 when on March 9, 1957, at 1422 GMT a major earthquake again shook the sea floor near the Aleutian Trench. At 4:22 A.M. local time in the fishing village of Unalaska, observers at the tide station were awakened by the shock. Eight minutes later, at about 4:30 A.M., the seismograph at the Honolulu Observatory registered the quake. The seismic alarms were set off, including the alarm in the nearby home of seismologist Harold Krivoy, who jumped out of bed and rushed to the observatory. Messages were immediately sent out to the observatories in Tucson, Arizona, and Sitka, Alaska, requesting additional information. When the reports came back in, the earthquake's epicenter was located at 51.3° north latitude, 175.8° west longitude, south of the Andreanof Islands. The magnitude measured 8.3 on the Richter scale—definitely an earthquake capable of causing a destructive tsunami. If a tsunami were on the way, it would reach Hawai'i between 8 and 9 A.M. that morning. A tsunami watch was put into effect.

Coastal communities in the Aleutians were contacted over military communication lines and requested to keep a close watch on the tides. At 6:50 A.M., Unalaska reported that the level of the sea had begun to rise. At a big sheep ranch on the northwest coast of the island, the waves surged ashore to within 25 feet of the ranch house and many sheep were drowned. At 6:53 A.M., Adak reported that waves 8 feet above normal had begun to break along the shoreline. At Sand Bay the tsunami surged in over 13 feet high and carried away a fuel dock. At Atka, the waves entered the harbor, destroying houses and boats and washing small boats up a creek. Near the Scotch Cap lighthouse on Unimak Island, a 40-foot wave was reported, and on the Pacific side of Umnak Island the waves were reportedly up to 75 feet high, washing driftwood a quarter of a mile inland. A tsunami was on the way across the Pacific, and it was time for a tsunami warning.

The warning system immediately alerted the police and military authorities. The police notified Civil Defense headquarters, the Fire Department, and the Honolulu City-County Emergency Hospital. Meanwhile, the warning system was having messages radioed to police and Civil Defense on the neighbor islands. Commercial radio stations began to broadcast the tsunami warning.

On all the Hawaiian Islands, police and firemen were clearing beaches and warning residents of low-lying areas and drivers on coastal highways. Fire stations and ambulance crews began to prepare for rescue operations.

Army and Marine helicopters were readied for flight. The Coast Guard sent messages to harbormasters throughout the islands to warn vessels to put to sea, and both U.S. Navy and Coast Guard vessels themselves began to head offshore.

At 8:44 A.M., the warning center received a message that the first waves had struck Midway at 7:45 A.M. that morning. In little more than an hour the tsunami would reach the main Hawaiian islands.

Kaua'i

Just before 9 A.M. the first waves began to strike the most northern of the main Hawaiian islands, Kaua'i. The 1946 tsunami had come ashore on Kaua'i like huge surf breaking, but in 1957 the tsunami surged in like a giant flood tide. At Hā'ena, on the northern coast of Kaua'i facing directly toward the Aleutians, the sea rose more than 32 feet above normal. Of the 29 homes in the small community, only 4 were left standing after the waves.

The villages of Wainiha and Kalihi Wai were virtually wiped out as the waves surged ashore on Kaua'i. Along with Hā'ena, they would be isolated for days because the bridges and roads connecting them to the rest of the island were destroyed.

In port at Nāwiliwili Harbor, the submarine U.S.S. *Wahoo* was awaiting a "small" disturbance from the tsunami. When the harbor began to alternately drain and refill, the skipper decided that it was time to get his vessel out to sea. While attempting to leave the harbor at a speed of 16 miles per hour, the *Wahoo* was actually carried backward by the strong tsunami current.

Along the beachfront of quiet, scenic Hanalei Bay, many homes were destroyed as the tsunami surged over the reef into the bay and up the normally tranquil river. Norman Kawamoto and Yohio Miyashiro were sitting in their small boat crabbing when the tsunami advanced up the river and capsized them. They clung to their boat for dear life as they were washed back and forth by the subsequent waves. Finally after 3 hours, they managed to grab hold of some hau tree branches. Here they stayed until they were rescued by Joseph Nakamura in a small boat. Fortunately, they were the only Kaua'i residents caught by the sea.

In all between 75 and 80 homes were destroyed or badly damaged on Kaua'i, more than twice the damage from the 1946 tsunami. Marine Corps helicopter crews worked around the clock to provision the areas of the island isolated by the waves, and the Hawai'i National Guard stood on guard against looting.

O'ahu, Moloka'i, and Maui

Next the tsunami waves reached the north shore of O'ahu. As on Kaua'i, the tsunami came ashore like a rapidly rising tide that just kept coming and coming. From the polo grounds at Mokulē'ia to the famous surfing beach at Waimea Bay, the water surged ashore to a height of 23 feet. Farther to the east past Kahuku Point, the tsunami engulfed beachfront homes at Lā'ie.

At Mākaha Beach on the southwest-facing Wai'anae coast, a visitor to the islands was making home movies of the scenic shore. As he filmed his friend standing on a low sea cliff, suddenly the water rose more than 15 feet, stopping just at the feet of his awestricken companion. The waves had begun to bend around the island.

At nearby Pōka'ī Bay, more than 50 small boats and half a dozen yachts were smashed against the new breakwater by the surge of the tsunami waves. One boat was washed over 200 yards inland and left to rest in a kiawe patch.

In the Ala Wai Boat Harbor, next to Waikīkī Beach, strong currents surged in and out, breaking off pilings and docks. Large yachts were washed back and forth, dragging sections of pier with them.

The tsunami continued to move south, passing Moloka'i, where the taro crop on the eastern end of the island was inundated and ruined. At Kalaupapa, the site of the community for patients of Hansen's disease, the waves rushed ashore to heights of more than 14 feet, smashing the settlement's water pipeline.

At the port of Kahului, Maui, the water surged in and out, creating a powerful vortex from the enormous turbulence.

The Island of Hawai'i and Beyond

At 9:17 A.M. the first waves struck Hilo. As the tsunami surged into the bay, the breakwater was inundated. The water rose 10 feet above normal at Pier 1, flooding the wharf by 2 feet and causing extensive damage to cargo in the warehouse.

Figure 4.3 Sequence of photographs showing inundation of ocean front property at Lāʻie, Oʻahu, during the 1957 tsunami.

Figure 4.4 People searching for fish on the exposed reef near Hale‘iwa, O‘ahu, during the 1957 tsunami.

Figure 4.5 Automobile abandoned by its driver on Kamehameha Highway at Waialua Bay, O‘ahu, during the 1957 tsunami.

Figure 4.6 The first large wave breaking over the seawall at Waialua Bay, Oʻahu, during the 1957 tsunami.

Figure 4.7 Man narrowly escaping from a tsunami wave near Sandy Beach, Oʻahu, during the 1957 tsunami.

Figure 4.8 Aerial view of Kahului, Maui, during the 1957 tsunami. Note the whirlpools formed by water withdrawing from the harbor.

The waves continued to move into the bay, submerging Coconut Island by some 3 feet and again damaging the small foot bridge. As the water moved up the Wailoa Estuary, several fishing boats were washed ashore or overturned. Watching from a utility pole near the Wailoa Bridge was fifteen-year-old Al Inoue. His vivid though patchy memories of the 1946 wave made him curious rather than cautious. Most of the small craft had been moved outside the breakwater, but not all. One in particular, the *Starlight,* had engine trouble and would not start. It was chained to the pier, but as the water lifted it the chains snapped and the boat was carried upriver, smashing into the bridge. Al was amazed to see the boat hit the bridge on the ocean side and come out the other side in pieces. The high ground around the Wailoa bridge was left as an island in a shallow sea of the tsunami floodwater. Reeds Bay was submerged once again to a depth of 9 feet, as it had been in both 1946 and 1952. Two houses standing on the point (the present site of the Nihon Restaurant) were damaged, with household objects washed into Liliʻuokalani Pond.

The pressure of the tsunami surging over the Hilo sewage outfall caused a manhole cover in town near the base of Mamo Street to be blown off, flooding the street. Along the unprotected sections of the bayfront, buildings were badly damaged, but the main area of downtown Hilo was largely spared. The coastal barrier and parkway between the Wailuku and Wailoa rivers had served as an effective buffer for this tsunami.

In low-lying Keaukaha the water flooded the ground floors of more than a hundred homes but caused little major damage. In fact, the total damage to homes, businesses, and boats for the entire Hilo area would amount to a modest $150,000, though elsewhere in the state it reached nearly $5,000,000.

The tsunami waves continued to move south past Hilo to Cape Kumukahi, where logs and cane trash were washed more than 12 feet above sea level. The western side of the Big Island, facing away from the Aleutians, was the area least affected. However, a 5- to 6-foot wave was reported to have come ashore at Keauhou at about 7:00 that evening, some 10 hours after the first tsunami waves had struck the northeastern side of the island. According to some experts, this wave could represent the reflection of the tsunami from the east coast of Asia.

In fact, oscillations from most tsunamis may continue for several days after the first waves strike. As late as 30 hours after the tsunami first hit Hilo, small tidal bores could still be observed advancing up the Wailoa River.

The tsunami also struck the coast of California. At Point Lobos, a man and woman who were fishing were swept off rocks into the sea. The woman managed to cling to the rocks until she was saved, but the man had to swim more than 100 yards through "stormy waters" to the base of a cliff where he was finally rescued. At Shelter Island near San Diego, a family was asleep on their 50-foot yacht, the *Sea Star*, when they were jolted awake. An 82-foot Coast Guard cutter, the *4F*, had been picked up by the surge of the tsunami, breaking its 1½-inch mooring cables; it washed into the *Sea Star*, crushing its skiff.

The tsunami continued to travel across the Pacific Ocean. It came ashore in the Marquesas Islands of French Polynesia as a 10-foot surge and on the coast of Chile it was measured at over 6 feet high.

1957 versus 1946

The speed, travel time, wavelength, and distance from Hawai'i of the 1957 Aleutian tsunami were very similar to those of the terrible 1946 Aleutian tsunami. The waves had traveled over 2,440 miles in 4 hours and 55 minutes, at a speed of just under 500 miles per hour. In spite of these similarities, the effects of the two tsunamis were quite different. The submarine earthquake in 1957 was greater than that of 1946, yet the 1957 tsunami was much smaller and produced much less damage on the island of Hawai'i. Kaua'i, on the other hand, was more severely damaged in 1957 than in 1946.

During the 1957 tsunami, only the northeast coasts of the islands recorded large wave heights, and these averaged only about 10 feet above normal, except in small, open bays such as Pololū Valley, where the water rose a prodigious 32 feet. In such areas, both the above-water topography and the submarine bathymetry (underwater topography) have an amplifying effect on the wave heights.

The complex interaction of the tsunami waves with the bathymetry was also well illustrated in Hilo Bay. Eyewitnesses reported that the third or fourth waves were the largest on the west side of the bay, whereas the tide gauge located on the east side showed the third and fourth waves to have been the smallest of the first four waves. A seiche of the water in Hilo Bay is one possible explanation. As water sloshed back and forth across the bay, the waves on one side could have been amplified, while those on the other side could have been diminished by the seiche. Observers at Coconut Island confirmed having seen a complex wave interference pattern produced near the middle of the bay.

Warning System

The warning system had once again proved its worth, and not a single life was lost to the sea as a direct result of the 1957 tsunami. On the Big Island in Hilo, most people (though not all) were kept away from danger areas long before and after the large waves.

On the island of Kaua'i there was much confusion on the part of the local authorities, in spite of the well-planned advisories from the warning system. Yet fortunately, even there, no lives were lost.

There were, however, two deaths indirectly related to the tsunami. A local reporter and photographer had chartered an airplane in order to try to observe and photograph the arrival of the tsunami. The small plane crashed into the sea off O'ahu, killing the pilot and reporter and injuring the photographer.

What had been learned from the tsunamis of 1952 and 1957? One fact stands out above all others: *each tsunami is unique.* The two Aleutian tsunamis of 1946 and 1957 showed that even tsunamis coming from the same general place of origin may differ greatly in their comparative severity at any one site. This was well illustrated on the islands of Kaua'i and Hawai'i, where the pattern of severity was reversed between the two tsunamis.

The probable size of a tsunami in Hawai'i had been predicted from reported wave heights at other places and from knowledge of past events. The specialists now began to see that many other factors must also be taken into account. The orientation of the Hawaiian Ridge and the coastline with respect to the direction of approach of a tsunami play an important role in determining the average water heights along Hawaiian shores. Local topography and bathymetry, for instance, small funnel-shaped bays, greatly amplified the heights of the waves, whereas reefs provided the most effective screen.

In Hilo the buffer zone provided by the coastal highway and park expanse had proved an effective shield by protecting much of the town from the moderate Aleutian tsunami. Would the barrier prove sufficient for a larger tsunami or a tsunami originating in South America? Many questions remained to be answered.

Evaluation of the danger to Hawai'i is based on information from tide stations closer to the source of the tsunami. During the 1952 and 1957 tsunamis, the warning system had admirably fulfilled the major function of reporting and evaluating the tide records and observations of water heights from the distant stations. Reports from the Aleutians and Alaskan areas and from Midway Island provided invaluable data to the warning center. In general, when a tsunami is generated in the North Pacific, stations in these areas, as well as those in Japan, Russia, Canada, and the U.S. mainland, may all provide information on which to base a decision about danger to Hawai'i.

For tsunamis coming from areas south of Hawai'i, the situation was not as clear-cut. In the case of an earthquake in Peru or Chile, there was much less information available and it was less dependable. A great deal of uncertainty was involved in trying to decide whether or not a destructive tsunami was headed for Hawai‘i. The warning system had always erred on the side of caution in such cases. However, this situation is not always appreciated by the public. A Mexican earthquake in 1958 resulted in an alert. No tsunami was generated and the public considered the warning as a false alarm. A well-located tide station could have immediately canceled the alert, but none existed. The warning system recognized the immediate need for more tide-observation stations between vulnerable areas and the probable sources of tsunamis, if unnecessary alerts were to be avoided. Yet at the time, resources were just not available.

In 1960 the South Pacific was still without sufficient outlying tide stations. In the event that a tsunami originated along the coast of South America, confirmed advance warning would not be available. The situation was prophetically stated in a 1959 scientific paper on tsunamis: "Hilo will be the outstation, and the only present recourse is a general warning without corroborative tidegauge data." How would the population react to another tsunami alert? To another "false alarm"?

5

Disaster by Night

The 1960 Tsunami from Chile

Sunday, May 22, 1960, was a day of terror for the South American country of Chile. It had been foreshadowed a day earlier near the city of Concepción, located between the high, rugged Andes Mountains and the great ocean depths of the Peru-Chile Trench. Just after 6 A.M. on May 21, Concepción was struck by a major earthquake. Damage to the city and to nearby towns was extensive and the earthquake registered an impressive 7.5 on the Richter scale. The Honolulu Observatory issued a tsunami watch at 6:45 A.M. (local time in Chile), and beginning at about 9 A.M. small waves 8 to 12 inches high were registered on the tide gauge at Valparaíso, Chile. The small waves continued throughout the day and night, gradually decreasing. Following the earthquake, the Honolulu Observatory had begun monitoring tide gauges throughout the Pacific. No dangerous waves were noted, though a very small but measurable wave had been recorded in Hilo Bay. It soon became apparent that a Pacific-wide tsunami had not been generated and finally at 8:49 P.M. the alert was canceled. However, just as a precaution the Chilean Navy ordered a permanent watch put on the tide gauge at Valparaíso. The Chileans were justified in their prudence, as much worse was to follow.

At 3:10 the next afternoon (1910 GMT, May 22), the shaking began once again with another 7.5 earthquake, but this was just the warmup. Only seconds later, as walls were still crumbling from the tremor, an even larger quake began at the same epicenter off the coast between Concepción and Chiloé (offshore from the province of Llanquihue). For the next 4 minutes the quake intensified as movement was extended along the fault in the earth's crust. For an instant the shaking seemed to slacken somewhat, but then it increased with unbelievable force. Finally, after nearly 7 minutes, it stopped. At its maximum, around 3:15 P.M., it measured a colossal 8.6 on the Richter scale—more than 30 times the energy of the earlier 7.5 quakes.

All of narrow southern Chile was severely shaken by the immense quake. In the city of Valdivia, nearly every brick wall fell and most wooden frame houses were broken or displaced. Double-locked, solid wooden doors flew open. Even cement walls were flung down and steel-reinforced concrete walls severely cracked. Parts of Chile suddenly dropped and others instantly rose. An area 20 miles wide along the coast from the southern Arauco Peninsula to southern Chiloé sank 5 feet. In Valdivia alone, more than 10,000 acres of meadows and farmland subsided and were quickly flooded. Wolfgang Weischert of the Universidad Austral de Chile watched in fascination as the Calle-Calle River began flowing upstream "with a greater velocity than I had ever seen the water flowing downstream." Just off the coast, Guafo Island rose 10 feet, suddenly connecting two formerly separate islets with a strip of land.

Meanwhile the seismic waves were traveling through the earth to the Honolulu Observatory, where they triggered alarms on the seismographs at 9:38 A.M. Hawaiian time (3:38 P.M. in Chile). From the time that the first earthquakes from Chile were registered at the Honolulu

Figure 5.1 Destruction from the 1960 earthquake at Valdivia, Chile.

Observatory the day before, the scientists at the warning center had remained in a tense state of alert. After the earthquake measuring 8.6 was recorded, authorities predicted that a tsunami had been generated—perhaps even a large, destructive tsunami. If a tsunami were on its way, it would take nearly 15 hours for the waves to travel the 6,600 miles between Concepción and Hilo; the tsunami would arrive, therefore, around midnight Hawai'i time. But in Chile the effects of the tsunami were already being felt.

At the southern end of the island of Chiloé, on a hill 200 feet above the sea, stood the Punta Corona lighthouse. When the earthquake began, the chief of the lighthouse, Señor Gabriel Jimenez, had evacuated his family and the rest of the personnel from the building and headed westward down the hill. At the height of the earthquake, they looked back to see the lighthouse tower tumble to the ground. Eight to 10 minutes later they froze in their tracks as they watched the sea in front of them begin to withdraw. After about 10 minutes, when more than 1,500 feet of sea floor had been exposed, they witnessed an enormous wave, from 50 to 65 feet high, form about 2,500 feet offshore. The wave seemed to be somehow held back as if waiting to gain strength and had smaller waves "boiling in front of it." Then this "veritable wall of water" began moving toward shore at great speed. It washed over the nearby Isthmus of Yuste, carrying away everything in its path, including small ships and houses. The tsunami would reach to a height of more than 70 feet near Quellon on the extreme southeast end of the island.

Meanwhile on the north end of Chiloé Island at the port of Ancud, the water had risen a mere 3 feet about 20 minutes after the earthquake, but then it began to withdraw forcefully, carrying a sailing ship with it from the harbor toward the open sea. When the withdrawal stopped about 30 minutes later, with the sea floor exposed to a depth of more than 15 feet, a second wave, this time an immense one, formed at the entrance to the bay and began to come ashore as a 50-foot wall of water. The master of the sailing vessel said that the tsunami wave was as high as the mast on his ship. Fortunately the vessel was far enough offshore to make it safely through the approaching wave.

Farther up the coast at Caleta Mansa ("Quiet Harbor," ironically), the motor vessel *Isabella* had been taking on cargo at the time of the earthquake. When the water first began to rise, the fast-thinking captain had

Figure 5.2 Aerial view of Isla Chiloé, Chile, showing tsunami damage and extent of wave run-up.

ordered the engines started and the mooring lines cut. The stevedores aboard stayed where they were, while those on the dock ran to safety on a nearby hill. The vessel headed out of port immediately and was far enough offshore to survive the 25-foot tsunami wave, which struck 15 minutes after the earthquake. This wave was followed at 10- to 15-minute intervals by progressively larger waves, 33 feet and 40 feet in height. At the entrance to the bay they were described as an "avalanche of water without breakers." Thanks to the quick action by her captain and crew and no small amount of luck, the *Isabella* survived with no loss of life. Other vessels along the coast of Chile would not be so fortunate.

Corral

The most detailed accounts of the terrifying events of the tsunami in Chile come from the Bay and Port of Corral at the mouth of the River Valdivia. Here the three large merchant ships *Canelos, Carlos Haverbeck,* and *Santiago* were anchored in the port at the time of the earthquake.

The *Santiago*, the first to enter the harbor, was already tied up at the dock and very securely moored there, with five mooring lines and two anchors set. The *Canelos* was not yet at the dock but moored in mid-harbor with two lines to a buoy and two anchors set. The *Carlos Haverbeck*, which had entered the harbor less than two hours earlier, was anchored farthest out in the bay, waiting to tie up to a mooring buoy.

The earthquake caused little destruction on shore at Corral but was strongly felt as a seaquake on board all three ships. The crews reported that the ships "bucked," masts and rigging swayed, and aboard the *Canelos* cargo shifted. The sea around the ships began "to bubble exactly as if the water were boiling." Small horizontal movements of the water caused the *Santiago* to tug at her mooring lines and the other two ships to bob around their anchors.

First Wave

After about 10 minutes the sea began to slowly withdraw from the bay, and then at 3:25 P.M., the bay began to refill, first slowly, then with increasing speed. Whitecaps began to form as the velocity of the water increased and continued to rise, flooding the docks and reaching the first rows of homes in Corral. The surge peaked about 15 feet above sea level and then began slowly to withdraw. The *Canelos*, moored to the buoys and with engines running, managed to resist the surge of the first wave, but the *Carlos Haverbeck*, farther out in the bay where the current was stronger, lost control of one anchor and began drifting toward the wharf where the *Santiago* was tied.

The *Santiago*, meanwhile, was trying desperately to free herself from the dock. This became more urgent as the *Carlos Haverbeck* was spotted bearing down on her. Unable to cast off in time, the *Santiago* braced for the impact of the *Carlos Haverbeck*. Fortunately a small boat was tied to the starboard side of the *Santiago*, and though totally destroyed, the little vessel softened the impact between the two ships. Just as the *Carlos Haverbeck* scraped past, the dock to which the *Santiago* was tied began to collapse. In the nick of time the remaining lines were cut and the *Santiago* was free.

Meanwhile, the *Carlos Haverbeck* was hurled over a buoy, with the buoy's mooring cables tangling in her propeller before the engines could be stopped. Now she was without power and totally helpless. Next the *Carlos Haverbeck* tried to pass mooring lines to the *Canelos*, which was still holding her position out in the harbor. The nearby tugboat *Puma* was called, and had just begun to carry lines between the two ships when the withdrawal after the first wave began. The current increased to a speed of about 10 knots. This proved too much for the mooring lines of the *Canelos*, which soon broke, and she was sucked out of the harbor, dragging both anchors behind her. With engines running, the crew began to weigh the anchors with hopes of heading out to sea. But the withdrawal was too swift, and the *Canelos* was soon left high and dry. This happened so quickly that before the engines could be stopped the propeller had spun in the bottom, throwing mud over the engine cooling condenser. Now her engines were dead, and the *Canelos*, like the *Carlos Haverbeck*, was at the mercy of the tsunami.

As if it were suddenly time for serious destruction, the tsunami withdrawal gained further strength. Many houses were now carried from shore out into the bay. The *Carlos Haverbeck* ran aground on a point of land. Surrounded by floating houses, terrified stevedores climbed down from the ship and began jumping from house to house, working their way toward shore. But before 2 minutes had passed, the *Carlos Haverbeck* began moving again, this time running aground stern first on another point. As soon as she was aground again, the crew attempted to put cables ashore hoping to escape. Meanwhile the second tsunami wave began to form offshore in front of the bay.

Second Wave

At 4:25 P.M., a 30-foot wave began to rush toward shore, described by onlookers as "[a] great wave of overpowering force that destroyed all that it encountered in its passage." The *Santiago*, although now free of the dock and under full power, was helplessly washed across the harbor, narrowly managing to avoid collision and grounding.

The tsunami seemed to pause for a moment at its peak and then began to withdraw. The speed of withdrawing water gradually increased to about 20 miles per hour, carrying with it all the buildings on land lower than 30 feet above sea level.

During this withdrawal two of the strangest phenomena of the entire tsunami were observed at opposite sides of the bay. In the water just off Corral, a bizarre depression formed in the surface of the water. Witnesses on the deck of the *Carlos Haverbeck* and on the surrounding hills described it "as if the water fell into a ditch 25 feet deep and 30 feet wide." Two tugboats, the *Pacifico* and the *Puma*, attempted to run upriver away from the "ditch" but were inexorably sucked toward it. The *Pacifico* was pulled in stern first, did a double somersault, and disappeared under the churning water. The captain of the *Puma*, however, turned his vessel toward the ditch at the very moment the *Pacifico* was tumbling in. The brave little *Puma* literally flew into the ditch at full speed, and then popped out the opposite side. The remains of the *Pacifico* were never found, but her captain was miraculously washed up alive on to a beach half a mile away. His last memory before losing consciousness was of his boat capsizing and falling into what he described as the "abyss."

Meanwhile, on the opposite side of the bay, a 15-foot waterfall formed as the tsunami withdrawal passed over a sand bank. Witnesses said that both strange phenomena, the waterfall and ditch, lasted for about 20 minutes, while the second tsunami wave withdrew.

Figure 5.3 Sketch of the water "ditch" and the tugboat *Pacifico* being drawn into the "abyss."

Also during the withdrawal, the *Carlos Haverbeck* was carried over a sand bank, running aground yet again. She stayed grounded for 20 minutes as "great whirlpools formed all around the ship." Without any orders, a dozen stevedores aboard took one of the lifeboats and quickly left the ship. After they had successfully reached shore, the crew then launched the other lifeboat and, with 22 seamen aboard, headed toward dry land. No sooner had they left the ship than the current strengthened and began to pull them out to sea. They immediately headed back to the *Carlos Haverbeck* and had begun climbing aboard, when the lifeboat started to break up. All but three, the third mate, the boatman, and one sailor, managed to clamber up the side of the ship. The boatman and the sailor disappeared into the water, but the third mate somehow managed to resist the current until the water was shallow enough for him to walk on the bottom. He immediately headed for the stern, hoping to grab hold of the propeller and avoid being dragged out to sea. But as soon as he reached the stern, he saw that the propeller blades had been folded double and were hopelessly tangled with the buoy cable. He then lost his footing and was carried out toward the open sea. In the middle of the bay, he was washed near a boat and climbed aboard, only to have the boat capsize moments later.

Third Wave

The third wave, which had been slowly building up offshore, now began to move into the bay. It gradually gained speed until it was traveling at over 20 miles per hour. The pilot of the *Canelos*, Señor Weller, said that at this time he looked toward the horizon and saw a wave whose height he estimated at 30 feet, "breaking and advancing with great velocity." As the second wave had cleared away most of the buildings on shore, the third wave penetrated even farther onto the land. It carried many of the houses washed out by the second wave and piled them against the hillsides.

The third wave now began to move all three ships. The *Carlos Haverbeck* was once again picked up and dragged over another sand bank. The *Canelos* was carried off, listing more than 45°. Part of the wave broke over the deck, leaving it full of sand, small stones, and fish. Then the wave picked up the *Canelos* and carried her toward the center of the bay at about 20 knots. She passed within about 500 feet of the lighthouse and was

Figure 5.4 Tsunami destruction at the town of Corral Bajo, Chile, 1960.

dragged over rocks and the breakwater at the mouth of the Valdivia River. In the process her keel was destroyed, and she began leaking fuel from her punctured tanks. While she was passing farther into the bay on one side, the *Santiago* was swept out of the bay on the opposite side.

Meanwhile the third wave picked up the third mate of the *Carlos Haverbeck* and carried him toward shore as he held to a floating spar for dear life. He would later be found alive on the beach 4 miles away. Unconscious, with a broken skull and two broken ribs, he was saved by his life jacket and incredible good fortune.

As the *Santiago* was being carried out to sea, she was surrounded by houses washed out of Corral. These included the house of the Corral harbormaster and the Corral firehouse. In the tower of the floating firehouse, its keeper could be heard still ringing the alarm bell. This brave man had gone into the tower to sound the alarm following the earthquake, and he continued faithfully at his post, ringing the bell, even though the firehouse was now floating in the middle of the bay. It was perhaps his karma, for he too would miraculously survive.

After about 10 minutes into the third wave, the *Canelos* was again

Figure 5.5 The merchant ship *Canelos* aground in the Valdivia River following the 1960 tsunami in Chile.

floated by the current and carried about 1½ miles upriver to sink into the mud in what would become her permanent resting place. Each following tsunami wave piled sand and mud against her side, increasing her list. Finally, fearing she would overturn, her crew finally abandoned the ship at about 6 P.M.

The *Santiago* survived the third wave, but was taken by the withdrawal, carried seaward with anchors dragging, and finally came to rest in front of a small peninsula. Here she would stay throughout the remainder of the tsunami.

Meanwhile back on the *Carlos Haverbeck*, the crew were still trying desperately to get off the ship and safely to shore. All the lifeboats were gone, so the first mate began ferrying the crew ashore in their tiny launch, 10 to 12 crewmen per trip. Small tsunami waves continued to wash in and out of the bay. During each trip they had to negotiate dangerous and unpredictable currents produced by these latter tsunami waves. Finally, at 10:45 P.M., the first mate returned to the *Carlos Haverbeck* for one last group, the captain and seven remaining officers. At the very moment the mate pulled alongside, a terrific current seized the launch and smashed it against the *Carlos Haverbeck*. The men in the launch were saved, but the boat was lost.

Now the *Carlos Haverbeck* began to list dangerously until she was tilted more than 60°. When the water reached the boat deck, the desperate officers decided to risk escape by launching the only remaining craft, a small life raft. At 11 P.M. the ship's officers climbed into the raft and finally abandoned the *Carlos Haverbeck*. The raft was immediately swept out past Isla Mancera, but before the men could paddle ashore there, they were washed back toward the inside of the bay. Next they were carried out toward the open sea and then north, passing within 600 feet of the *Santiago*. They signaled the *Santiago* for help, but her crew was unable to render aid as they were desperately trying to save their own ship.

The raft continued to be washed around the bay until finally at about 1:30 A.M. (May 23), they were left stranded on a beach south of the entrance to the bay. They quickly climbed a cliff to a house where they spent the rest of the night. Strewn around this house, at an elevation of over 80 feet above sea level, they saw the remains of other houses and debris thrown up over the cliff by the tsunami. Returning to Corral later that morning, they walked past the mooring barge from the port and the second story of a local hotel, both lying in a canyon behind a eucalyptus forest.

The *Canelos* and the *Carlos Haverbeck* were total losses. The *Santiago* had amazingly survived the tsunami—but the tsunami was not yet finished with her. On the morning of May 23, she weighed anchor and began to search the area in front of the harbor in hopes of rescuing any survivors. Later in the day, she headed toward Valparaíso, but while en route abnormal currents created by the tsunami caused her to steam off course, and she ran aground on Isla Mocha. The *Santiago* in her turn had succumbed to the tsunami's power. All three of the ships that had been at Corral were lost to the tsunami.

Further along the coast at Mehuin, the harbormaster reported the "tragedy of the fishermen." Observers on a hillside watched in horror as a group of fishermen with their wives and children tried to flee from the tsunami. Running through thick brush, where they repeatedly fell, many were caught by the surging water and drowned. The nearby Marine Biology

Figure 5.6 Aerial view of the merchant ship *Carlos Haverbeck* submerged off Corral, Chile, following the 1960 tsunami.

Station was carried out to sea by the first wave and then washed back to shore, where it was smashed against coastal cliffs.

Virtually every port along the coast of Chile, from Concepción to Chiloé, had been devastated by the tsunami. The death toll in Chile from the earthquake and tsunami would eventually reach 1,000 and the damage would amount to a staggering $417 million. But the tsunami had just begun on its path of destruction.

Following the earthquake, information about the tsunami had gradually begun to trickle in to the Chilean Office of Navigation and Hydrography in Valparaíso, the local headquarters for the tsunami warning system. They were now prepared to send a cable to the Honolulu Observatory confirming the generation of a tsunami.

But meanwhile the Honolulu Observatory had already issued the following bulletin:

> This bulletin is a tidal wave alert. A violent earthquake has occurred in Chile. . . . It is possible that it has generated a large tsunami. Although we have as yet no data, we are awaiting information from Valparaiso and

> from Balboa. If a tidal wave has been originated it should arrive about midnight Hawaiian time today at the Island of Hawaii and 30 minutes later at the Island of Oahu. New information will be given as soon as more data is available.

Five minutes later the message from Valparaíso arrived, confirming that at least a local, if not a Pacific-wide, tsunami had been generated. An hour and a half later the warning center issued another bulletin indicating that a tsunami "might" be crossing the Pacific, but still no firm official warning was issued.

By this time the tsunami waves had already begun to reach the more distant islands off the coast of Chile. The residents of the Juan Fernandez Islands, 400 miles off the Chilean coast, were fortunate. The vast majority of the population of the main island was on the west side, away from the tsunami's path of approach, watching the arrival of the sailboat *Robinson Crusoe,* coming in from Valparaíso. Though the sea withdrew about 100 feet, only small waves came ashore, and these did little damage.

The tsunami, continuing on its path across the Pacific, next struck Easter Island. Here, too, the population was little affected by the waves. Most of the permanent residents of Easter Island live near Hanga Roa, on the west side of the island. On the east side, facing South America, waves 20 feet high washed ashore, penetrating about 1,600 feet inland. Windmills that operated water pumps were destroyed, as were some stone fences. Most significantly the tsunami destroyed the Ahu of Tongariki. Ahus are stone platforms on which the islanders erected the stone statues for which the island is now so renowned. The Reverend Padre Sebastian Englert, who visited the site shortly after the tsunami, described the scene:

> I saw with surprise what may be described only by the graphical expression that "no stone remained upon stone"! An enormous avalanche of water, probably some 6 meters [20 feet] over the level of the sea, invaded the coast of Hotuiti for a distance of 500 meters and swept away the ahu, statues, sunshades, fences, and blocks of stone with such force that only a person who had known the ahu well before its disappearance could now establish its former location.
>
> The waves of the sea appear to have played with the heavy statues as if they were balls. They remain scattered around 50 to 150 meters from the site of the ancient ahu. I was especially taken by one of these and was

> struck by its beauty. . . . I had not appreciated the state of sculptural perfection because the head and chest had been partly buried. Now it lay stretched out mouth upward among stones and rubble some 100 meters from its former location. . . . it weighs at least 20 metric tons. The waves of the sea transported it with an irresistible, but at the same time such gentle, force such that it had not deteriorated nor fractured, and the perfect form of the face and chest cause admiration.

Though destroying the platform, the tsunami had uncovered one of the most beautiful of the famous statues of Easter Island.

Meanwhile back in Hawai'i, reports from the news media began to drift in from Chile, telling of destructive waves and damage along the coast of South America. The scientists at the Tsunami Warning Center knew that a decision about a tsunami warning had to be made—a decision without the backing of reports from stations along the path of the waves between South America and Hawai'i. Preferring to err on the side of caution, the warning center issued a tsunami warning at 6:47 P.M. However, only 40 minutes after they issued the official warning, fresh doubts were raised as they received messages from Balboa and Christmas Island stating that no tsunami was observed.

In spite of these doubts, the center continued to send warning messages across the Pacific. In American Samoa, warnings were received at 4 P.M., 6:11 P.M., and 8:08 P.M., with the arrival time at Tutuila predicted for 10 P.M. Western Samoa, however, would receive no warning messages from the center in Honolulu. They had inadvertently been deleted from the list for warning notification. Fortunately, the duty officer at the local warning center in Fiji, on his own initiative, relayed a message to Western Samoa and the residents were not caught completely unaware of the danger.

At 8:30 P.M. the coastal sirens in the Hilo area began to sound. Then just after 9 P.M., radio stations carried reports from Tahiti stating that the tsunami had reached Papeete and that the waves were an unspectacular 3 feet high. Was this tsunami warning to be seen as yet another false alarm? How would the population react to the warning now? What most people didn't understand was that except for the Marquesas Islands, most of French Polynesia is relatively safe from tsunamis. The islands are surrounded by well-developed reefs with steep offshore slopes. These factors work to minimize the effects of tsunamis.

The warning center in Honolulu finally received an official message from Tahiti at 10:23 P.M. indicating that unusual wave activity had occurred. Amazingly, this was the first official confirmation of the tsunami, other than from the coast of Chile. Though the tsunami registered only about 3 feet in the harbor at Papeete, there was local inundation of up to 10 feet observed at the Faratea Hotel and Restaurant, where the ocean-side bar was completely destroyed. Fortunately, the bartender heard the warning over the local radio and cleared the bar before the arrival of the tsunami.

In contrast to the Society Islands (Tahiti), the Tuamotus, and the rest of French Polynesia, the Marquesas Islands have few outer reefs and a more gradual bottom slope. Large deep bays with gradually rising bottoms can cause waves to build up to great heights. Furthermore, the villages are mostly situated on lowlands at the heads of these bays and as a result are very vulnerable to tsunami damage. Local witnesses estimated that the largest wave was as high as 30 feet on the headland under the lighthouse at Taa Huku. At Atuona several buildings were destroyed, including the jail, the radio station, and a church. Fortunately the only life lost was that of a horse that had been left tied to a coconut tree.

Just before 11 P.M. the tsunami reached Samoa. As in most of French Polynesia, here the coast is surrounded by coral reefs, and the tsunami mostly consisted of a series of rapid surges with no big waves. But in bays with a gradually sloping sea floor such as at Pago Pago in American Samoa and Fagaloa Bay in Western Samoa, the tsunami waves reached their maximum height. At Pago Pago village a house was lifted from its foundation and washed into another, and at Fagaloa Bay the tsunami produced an 8-foot wave, which advanced 90 yards through the village, reaching the roof of a native house. Fortunately, no one was sleeping in the house at the time and no lives were lost in either American or Western Samoa, though damage was estimated at about $50,000.

Impact on Hawai‘i

In Hilo most people had heard the news that a tsunami was supposed to arrive about midnight. But many people didn't really understand the warning. In fact, just a few months before, the system of warning sirens had been changed. Under the old system there were three separate alarms: the first siren indicated a tsunami warning was in effect; the second meant that it was time to evacuate; and the third was set to go off just prior to the arrival of the first waves. Under the new system, there was only one siren—and it meant "evacuate immediately." After hearing the "first" siren, many people began to pack up their belongings in preparation for evacuation. Then they waited for the second siren before leaving their homes. There was to be no second siren that night!

The reaction of Hilo residents varied greatly, even among those who had firsthand experience of the wave in 1946. Some took it very seriously indeed. In Keaukaha, most of those who had seen the destruction of 1946 evacuated immediately. Chick Auld recalled how the police went from house to house, making sure that everyone was aware of the threat. Chick and his family left—so too did the Cooks and the Talletts. Paul Tallett wondered for a while if it was really necessary, because there had been false alarms since 1948, but in the end he decided to be cautious. Similarly, the staff at the Hilo Yacht Club left their workplace, wondering whether once again the building would be washed away. This time they took all the important documents with them. All through Hilo, businessmen were moving documents and money from their vulnerable offices to safer ground. Although Bobby Fujimoto had relocated his planing mill uphill of Kamehameha Avenue after 1946, he decided it would be wise to remove his books and accounts. Many others took the same prudent action, but not everyone. Some were undecided. It was not that time had dimmed their memories of the shocking experience of 1946—they remembered with a fixed and awful clarity. But they could not believe that it was going to happen all over again, that this warning might need to be heeded when many in the past had seemed a waste of time.

Someone else with vivid memories of 1946 was "Baby Dan" Nathaniel. In 1960 he was working as a tour guide and had a party of tourists at the Hawai'i Volcanoes National Park when the tsunami warning was given. He took his group back to their hotel—the old, wooden Naniloa—and told them to pack their bags. At 8 P.M. they were taken to the Hilo Hotel on Kino'ole Street. Working at the Naniloa was Martha van Gieson, who had run from the tsunami of 1946 in Keaukaha. Despite her firsthand knowledge of the ocean's strength, she remained at the hotel with her colleagues, partly to protect the property from theft but also to make sure that all the guests had gone. When the staff checked, they discovered 10 people waiting in their rooms to photograph the event, and persuaded them to leave.

Geologists at the Hawaiian Volcano Observatory, aware of the earthquake, had waited anxiously throughout the afternoon, listening to the radio for further news. Just before 10 P.M., a group of them drove down to Hilo from the national park. In spite of the darkness, they felt that they might be able to make useful observations of the tsunami. After arriving in Hilo and clearing their plans with the local police, they drove through the now-deserted streets to the Wailuku River. There they set up their observation post on the bridge overlooking Hilo Bay—the same bridge where just over 14 years before Jim and Bob Herkes had witnessed the awesome waves of the 1946 tsunami. The geologists kept busy measuring the heights above sea level of various reference points on the bridge pier opposite them. They intended to document the water level of each wave of the tsunami. They also planned their own evacuation route, a short sprint along the highway to safe, high ground. And when all was noted and ready, they waited.

The relatively small tsunamis of 1952 and 1957 had given some residents a false sense of security. The curious, the foolhardy, and the misinformed actually went down to the bay to wait for the waves to come in. They stood around the old Suisan Fish Market filled with the excitement and sense of adventure instilled by a late-night outing. All of Hilo waited for midnight and the new day, May 23, 1960.

Just after midnight, the geologists observed that the water under the bridge had begun to rise. Within 5 minutes it reached a crest 4 feet above normal. Then the water slowly fell, and by 12:30 A.M. a trough 3 feet below normal was recorded. The first wave from the Chilean tsunami had arrived. A few minutes later, word reached the geologists that the first wave had flooded the sidewalk near the bridge across the Wailoa River at the south end of the bay.

Reports from a Honolulu radio station stated that no waves had yet arrived at Hilo and that the estimated arrival time had been set back half an hour. Yet with their own eyes, the geologists had seen the first wave pass. The radio report could only mean that communication between the warning system and the news media had broken down. Misinformation had now increased the danger to the public.

Figure 5.7 Curious Hilo residents await the first wave of the 1960 tsunami.

At 12:46 A.M. the second crest washed under the bridge at a level 9 feet above normal. The wave topped the seawall of downtown Hilo and flooded the heart of the business district along Kamehameha Avenue. This wave was as large as the largest wave of the 1957 Aleutian tsunami.

Now the water began to withdraw again from the bay. At 1 A.M. the level measured nearly 7 feet below normal. At this point a strange calm prevailed, which the geologists described:

> At first there was only the sound, a dull rumble like a distant train, that came from the darkness far out toward the mouth of the bay. By 1:02 A.M. all could hear the loudening roar as it came closer through the night. As our eyes searched for the source of the ominous noise, a pale wall of tumbling water, the broken crest of the third wave, was caught in the dim light thrown across the water by the lights of Hilo. It advanced southward nearly parallel to the coast north of Hilo and seemed to grow in height as it moved steadily toward the bayshore heart of the city. At 1:04 A.M. the 20-foot-high nearly vertical front of the in-rushing bore

Figure 5.8 Police officers and residents watching the second wave flood into the Wailoa area of Hilo.

churned past our lookout, and we ran a few hundred feet toward safer ground. Turning around, we saw a flood of water pouring up the estuary. The top of the incoming current caught in the steel-grid roadway of the south half of the bridge and sent a spray of water high into the air. Seconds later, brilliant blue-white electrical flashes from the north end of Kamehameha Avenue a few hundred yards south of where we waited signaled that the wave had crossed the sea wall and buffer zone and was washing into the town with crushing force. Flashes from electrical short circuits marked the impact of the wave as it moved swiftly southeastward along Kamehameha Avenue. Dull grating sounds from buildings ground together by the waves and sharp reports from snapped-off power poles emerged from the flooded city now left in darkness behind the destroying wave front. At 1:05 A.M. the wave reached the Hawaiian Electric (HELCO) power plant at the south end of the bay, and after a brief

Figure 5.9 The Waiākea town clock stopped at 1:05 A.M., when the third and largest wave struck. The clock now stands as a monument to those lost to the 1960 tsunami.

greenish electrical arc that lit up the sky above the plant, Hilo and most of the Island of Hawaii was plunged into darkness.

Men like Oliver Todd who loved the ocean—swam and fished in it—could not imagine living away from it. Even though his home had been destroyed in 1946, he and his family had returned to live on the oceanfront. Their new house had been built on land leased from the state (on the site of the present-day Hilo Hawaiian Hotel). On May 22, their son Oliver Jr. was home on leave from the navy. He and his father loaded some of their most valuable possessions in a truck but decided to stay at the house and await events. Josephine Todd went to Villafranca (the area around Hualālai Street) in the evening. Like most people in Hilo, they listened to the radio. However, it was not from the radio that the Todds learned that the wave was an actuality rather than a possibility. The warning note was sounded by empty oil drums banging together underneath their house. When the two Olivers heard the sound of the oil drums, they knew the water had risen behind them. In their heavy-laden truck they drove in the direction of Waiākea School but were soon confronted by a flood of water rushing toward them. Without hesitation, they took the only action available to them and swung the sturdy vehicle straight through the bushes into the houselots subdivision.

Many of the people living in that subdivision were to remember that Sunday in May as one of the worst days ever. The Sakai family started off on their normal family activities, with Mr. Sakai taking his sampan out into Hilo Bay to fish and Mrs. Sakai taking the children to Coconut Island to swim. When they returned home from swimming, Kimiko Sakai had a telephone call from a friend who worked at Suisan Fish Market to say that a tsunami warning had been issued. She sent one of her sons to signal his father to come back to shore. Then Mr. Sakai and his brother took all their boats out to sea. In the mean time, Mrs. Sakai prepared to evacuate but was unwilling to actually leave her home because she remembered that the waves in 1946 had not reached the house. She heard a siren, but she expected a policeman to come and tell her when to leave. Instead, after midnight, she heard the voice of her neighbor saying that her slippers were floating in the yard. When Kimiko rushed to the window and looked out to Coconut Island she saw water under the banyan trees reflecting the light of the street lamps. As she stared at this unexpected scene, her heart racing, her son came running back saying that there were mullet flapping in the bushes at the back of the house, where they had been left by the receding wave. She thought that the time to leave had certainly come, so she gathered up a few important papers and ran outside. Her neighbors shouted for them all to get in their car, and drove off toward the Waiākea bridge. A policeman was there directing traffic, and he shouted that they should turn around. They drove back along Kamehameha Avenue, to find themselves confronted by a great rush of water. A tremendous explosion filled the air as the wave hit the power plant. They jumped out of the car and ran to the Sure Save Supermarket building, where they helped each other onto the roof. They perched there, clinging to each other, and in the darkness that followed the explosion at the power plant they could hear the suck of the receding water, the cracking and rending of buildings, and over that the mooing of a cow, cries for help, mothers calling to their children. After a long time the firemen came to check if they were safe, and they made their way back to the house in ʻAʻalapuna Street. There was very little of it left on the site. Most had been pushed off its foundation and moved toward Reeds Bay. A beer truck had been washed into the bedroom wall. But although their possessions had gone, they were all safe, including Mr. Sakai, who returned at about the same time, having saved his sampan.

The men who took their boats out of the harbor that night had a very different experience of the tsunami from that of the people fleeing on land. They found themselves on top of the giant waves.

Frank Paulo and a crewman, Thomas Kahulamu, left the Wailoa River on the *Leilani* about fifteen minutes before midnight; because they had no motor they were towed by Frank's brother in the *Lanakila*. They waited about 2,500 feet offshore for half an hour and then heard a noise from the direction of the breakwater. They saw a white-capped wave approaching from the direction of Paukaʻa Point, and headed into it. As they rose up the front of the wave the towline snapped. The *Lanakila* headed out to sea, but *Leilani* was left at the mercy of the waves. Three more steep fronts came in what seemed a very short time, each wave carrying the *Leilani* toward the shore, its withdrawal taking the boat back out again. Finally Frank Paulo was able to rig a sail from a tarpaulin and sail eastward. They stayed out at sea until about 5:30 A.M., when they moored inside the breakwater and went to report to the Coast Guard. Later that morning when Frank went back to check his boat, he found an old man, Mr. Shimazu, who had been washed out into the harbor in the wreckage of his house, then cast up onto the breakwater. Frank put Mr. Shimazu on the *Leilani* and they were towed ashore.

Other boat owners were also assisting the Coast Guard in searching for survivors. One of these was M. C. Child of the Hilo Technical School. He had taken his sampan *Kilohana* out of Hilo Harbor before the arrival of the tsunami and had just gotten outside the entrance when two large waves arrived one after the other. A couple of minutes after they had passed, he heard popping as the waves brought down the utilility poles and shorted the lines, then saw the flashes as the water hit the power station. Like Frank Paulo he stayed out of the harbor all night.

These men and many others had made the decision to take their boats to safety. Back in the houselots subdivision Al Inoue had made a different choice. He had fragmentary memories of the disastrous 1946 waves, when he had been just four years old, but he also had more recent memories of the smaller 1957 ones that he had watched without coming to any harm. Now a senior in high school, when he first heard the warning he went to see his friend Jimmy Lee, who owned a liquor store on the corner of Manono Street and Kamehameha Avenue. Inoue asked Lee if he needed any help to move things from his store. But Jimmy Lee, like so many others, was confident that the wave would not affect him much. He moved his stock to the higher shelves and took Al to see the high water mark made by the 1946 tsunami in the back alley. Still more reassured by this, Al Inoue decided, as the predicted arrival time approached, to go down to the Wailoa River with a friend to see what would happen. They parked their cars on the corner of Piʻilani and Manono streets and went to watch upriver of the Wailoa Bridge. They had been encouraged in the belief that there would not be a big wave because they had heard the report that the wave had been measured at Tahiti at 3 feet. Even when the water started to behave in a strange way, Al remained unconcerned. The water in the river receded—but, he reasoned, the river was only half empty, not completely empty, so it could not be that an enormous wave was coming. The sound of exploding power lines from the north shore of Hilo Bay alerted him. Al saw those who had been watching from the Wailoa Bridge begin to run. Looking beyond the bridge, he was amazed to see a wave cresting at the mouth of the river, near the Suisan. It looked to him twice as high as the bridge. With his friend he turned and ran down Manono. Behind him he could hear the power lines whipping and crackling as the poles fell. He wanted to cross the road, but others who had been watching from their cars were now driving very fast up Manono. He did not know at that moment whether he would be drowned, electrocuted, or run over. But he was a lucky young man. He outran the water and reached his car, turning on the headlights to act as a beacon to others in the darkness.

Not quite so lucky was Takeo Hamamoto. His home was in Lili'uokalani Gardens, which he tended for the county. He had heard the warning and sent his son and daughter to safety with his brother. He stayed at the house with his wife and work crew (which included Taffy Okamura, who had memories of running from the 1946 wave), with the idea that they would save some of their possessions if the need should arise. Around midnight—the estimated wave arrival time—they saw by the light of the lamps on the bridge to Coconut Island that the water in the bay was receding. Then the water began to flow into the park, first covering the parking lot. Takeo decided to move his family's new washing machine to higher ground, so they took it to the ice plant; however, just as they had done so, they heard a thundering roar as the big wave hit the coast and saw white water breaking over Coconut Island. With all haste the Hamamotos climbed into their car, and Taffy Okamura and the rest of the crew piled into the pickup. They drove along the road (where the present golf course is) but found buildings blocking it. They turned toward the Naniloa, but found more collapsed buildings. The vehicles were caught by the wave. How could they save themselves now? With the water surging behind them, they ran into the Waiākea schoolyard, where the crew climbed into the trees and Takeo and his wife clung to a large monkeypod tree. Building debris washed all around them—the water was up to Takeo's chest. As the crest of the wave passed, he and his wife climbed higher into the monkeypod tree. Hearts hammering with the exertion and excitement, they had another fright as a bright blue flash and loud explosion signaled to all of Hilo that the wave had reached the electric power plant.

The explosion of the power plant was the first indication for Tom Okuyama that the big waves had arrived. Working in his office at the Sure Save Supermarket (then on Kamehameha Avenue), he had been listening

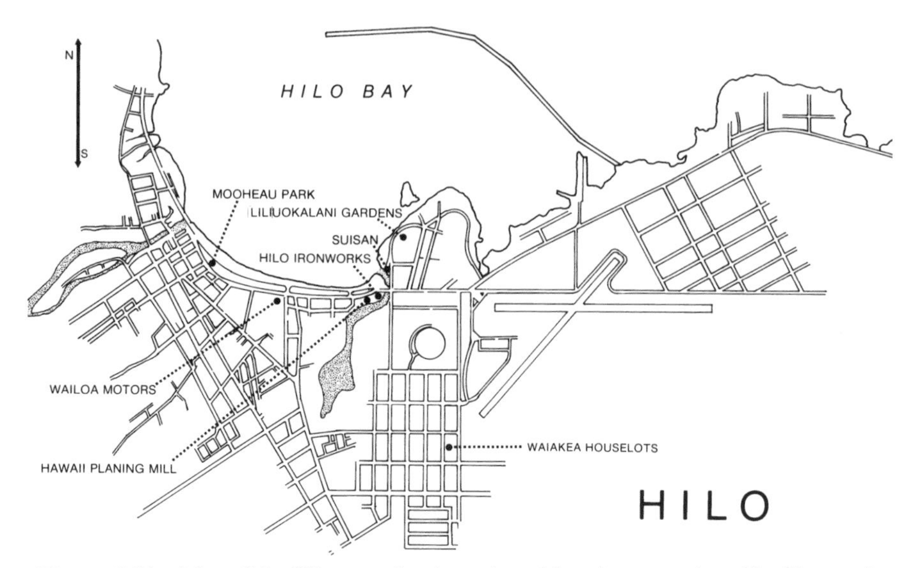

Figure 5.10 Map of the Hilo area showing selected locations mentioned in Chapter 5.

to the radio but heard no news of the first waves. The water inside the supermarket was knee-deep, but he was able to make his way out through the back of the building.

Martha van Gieson was driving away from the Naniloa, toward the Hilo Hotel. She and her colleagues had waited until after midnight, keeping in touch with Civil Defense. She phoned them to say that water had advanced about 7 feet up the lawn. They were in the lobby on the second floor, and decided to stay. Around 1 A.M. all the lights on the switchboard started to go on and off, so Martha guessed that the basement was flooded. Martha phoned Civil Defense to tell them, and was asked to stay on site to give reports. She decided her cooperation had gone far enough, and left, luckily for four boys who had foolishly gone to watch the wave; she picked them up and returned them to their homes.

Nearer to the Wailoa River, the force of the waves was greater, as Al Inoue had seen. Evelyn Miyashiro had not urged her husband to move away from the area, despite her experience in 1946, when a tsunami wave had washed over her on the bridge. The restaurant the Miyashiros owned had done well; indeed, just 23 days before this night they had opened new premises, this time on their own property. It was not surprising, therefore, that they wanted to stay near both their business and their new home next door, in which they had lived for just one year. After the warning had been issued, Evelyn and Richard decided to stay with their property, since the flooding there had not been too severe in 1946. Together they watched people tying up their boats at the riverside. The tension present in each one was released by the wild cries of those who had been staring out to sea: "A big one is coming!" The Miyashiros thought of trying to drive away but could not find their fourteen-year-old daughter. They ran to their home, where they found her praying upstairs. Then the whole family—mother, father, and three daughters—joined hands and prayed. As they prayed, they felt the house being lifted from its foundations and carried along on the water. The lights went out and they had no idea where they were being carried. Evelyn was sure they were being taken out to sea, and that they were about to die. For what seemed a very long time they stayed in the room holding hands, until they saw the light of a flashlight and heard a voice calling. The house had become wedged on a piece of higher ground. The Miyashiros felt themselves blessed in having been saved, even though their new restaurant was destroyed. For the second time the sea had attacked their business, this time more fiercely than before.

Tadayoshi and Hisako Okamoto's store on Mamo Street was also washed away by the big waves. They had repaired the store after the 1946 tsunami, but they moved from the apartments above the store to a house on Pi'op'io Street. The family was asleep in their home when the waves began. They had heard the warning earlier but thought it was just another false alarm. Mrs. Okamoto had been at the beach that afternoon with her children, and she observed that the sea was calm, with no hint of the wave traveling at high speed across the Pacific toward them. The first they knew of its arrival was a "roaring like an express train." Their reaction was to spring from bed and turn on the lights, but as soon as they had done so the lights were extinguished by the wave's assault on the power plant. Realizing that it was too late to run, Mr. Okamoto put a mattress over his wife and three children to protect them from debris should the walls collapse. But like the Miyashiros, they were among the lucky ones. When the biggest wave had passed, they waded through the flooded ground floor of the house to their backyard. With superhuman strength Mr. Okamoto was able to wrench open the heavy gate, and in their night clothes and bare feet the family fled the area and ran uphill as fast as they could.

Figure 5.11 Men trying to keep a fishing boat tied at her berth in the Wailoa River, Hilo, as the water withdraws before another tsunami wave.

Also on Pi‘opi‘o Street, the Kinoshita family had made the same decision not to evacuate. The 1946 tsunami had brought debris to their street, but nothing worse. So despite the warning issued earlier in the day, they confidently believed reports that the danger was over, and all went to bed. The mother and father went to sleep, but their children, Bert and Janet, stayed awake to see if anything would happen. More happened than they had ever anticipated. They heard an ominous rumbling, looked out of the window, and saw electric flashes light the sky as the water struck the power poles. Then all was dark, and along with all the other houses at the lower end of Pi‘op‘io Street, theirs was lifted off its foundations by the wave. When it was dropped to the ground, the water was chest high. They had no idea what would happen next, and decided that if the water rose more they would have to swim out the window. Mrs. Kinoshita could not swim; because she was short, Bert had to hold her above the water. She was faint with terror, but Janet slapped her face to revive her. They all held on to each other, fearful that this was the end but thinking that at least they would be found together. When the water started to recede, the house began to move again. They could see nothing in the awful blackness, and since the front of the house was damaged they could not get out. In the darkness they could hear their neighbor calling for his son. They waited for dawn, and rescue.

The wave had come from the Christina Lane side. On that street (now the site of Long's Drugs' downtown store) lived Everett Spencer, his wife Amy, and their daughter Lynette and sons Sheldon and Robin. Everett had been an eighth-grader at Laupāhoehoe High School in 1946. He had run from the wave across the ball field and he had seen the devastation that it wrought. But despite the tsunami warning, he did not expect the predicted wave to have a serious effect on Christina Lane. He went to sleep. He awoke with a jolt to find his roof gone, and he saw a bright blue light in the sky. There was a rumbling noise, and the house was moving. Amy thought the world was ending. When the house hit ground, Everett called for his children. For some reason they had decided to sleep together that night instead of in their own rooms. Sheldon responded to his father's calls with the news that Lynette was with him but he could not find his baby brother. Everett got his family out of the house and onto higher ground, and ran back to look for one-year-old Robin. He called and called and was finally rewarded with a whimper from a neighbor's house. There he found his son and pulled him from the debris, bleeding. He ran with him to the home of Dr. Francis Wong, whom he found sleeping, unaware of the disaster. When they arrived at Hilo Memorial Hospital, they were the first patients, and no one there knew of the tsunami either.

Another 1946 survivor lived nearby, on Kumu Street, next to the canal. June Odachi was now married to Sumu Shigemasa, and they had a twenty-month-old baby daughter, Junette. They had prepared for evacuation but waited to see whether there really would be a wave. They had heard that in Kona the wave measured only 1½ feet. Sumu, an enthusiastic fisherman, was very interested when the water receded and he saw fish flopping around in the canal, so he went to get his fishing line. June could hear a strange noise, but her husband told her she was imagining things. Then a Filipino man ran up the street, shouting, "Hai, the wave comes." (They were later to see his name listed among the dead.) They ran into the house, just as it was hit by the wave, which churned through the building and filled it with water. They rushed to the front and saw that the bedroom was gone, so they ran to the back bedroom and jumped onto the bed. The water was still rising and they balanced on the bed frame. Sumu went under the water to look for a way out, broke through the window screen, came back, and told June to swim out. He took the baby, explained to her what he was going to do, then covered her mouth and nose with his hand and took her under the water and out of the window. Junette did not swallow one drop of water. They climbed onto the roof, where Sumu tied June and the baby to the television antenna with the cable, while he went off to find something to float them on. Hearing neighbors calling for help, he went to them first. June soon untied herself, worried that if another wave came she might be dragged under with the antenna. After what seemed a long time, Sumu came back with some wall siding but it was too big and heavy. Then he found a plank, and took them to a mango tree. National Park rangers rescued them from it in the darkness and took them to the hospital.

On Mamo Street another family had evacuated from their home. Sixteen-year-old Carol Brown helped her father collect valuables, which they took to her sister's home in Pāpa'ikou. People passing in the street made fun of them because they were leaving. Carol and her brother Ernest went to another sister's home in Koa Lane, and borrowed her car to go and check on a niece who was babysitting. After they had talked to her, they decided that they would return to Hilo; they heard on the radio that the waves were 7 feet high. On the way they met a police officer, who told them that the danger was past, so they went back to Koa Lane and Carol began to prepare something to eat while her sister started to put the younger children to bed. Then came a rumbling, which they at first thought to be trucks; then it got louder, seemed closer, and was accompanied by crashing and crunching noises. Carol and her sister peered out from the top of the Dutch doors, until Ernest ran and slammed the door shut and shouted, "Stay in the house!" They followed him into the kitchen, just as the wave hit the building, floating it off its foundations. Carol felt the water rise and heard her sister screaming their names. She felt that she was going to die. Looking out of the window, she saw a chicken on the roof, and then her brother Ernest holding the roof's edge. He grabbed hold of a tree, which the house almost crashed into, then the house floated past him, inland. After the water receded, he shouted at them to get out of the house. They had to move the dining table away from the door and then jump down because the porch had gone. They were very happy to be safe, but concerned for their parents. When they found them on Haili Street, Carol wept tears of thankfulness and relief that her family was spared.

All over Hilo people were having lucky escapes, even though they had remained in the inundation zone. Mark Olds lived near the Todds, on the site now occupied by the Hilo Bay Hotel. He was one of the many residents who evacuated the danger zone only to return before the danger had passed. Although he heard a tsunami warning on the radio around 2 P.M., he did not leave his home until 4:30 P.M., when another warning was broadcast. In his office in the downtown district he continued to listen to his radio. As he remembers, he did not hear any statement of the expected time of arrival of the first waves. He did hear that the tsunami had reached Tahiti, where the height of the biggest waves was no more than 3 feet. On hearing this information, Mark decided to return to his house, where he spent the evening watching television. No warnings were broadcast by the television station. Around midnight he was ready to go to bed, but on impulse he thought he would take a look at the ocean before retiring. Opening his back door, he switched on a yard light, which revealed that the yard was completely flooded. Mark decided it was time to leave and moved quickly to the front of his house, which was at a higher elevation. As he reached the lanai, the explosion from the electric plant reinforced the urgency of his situation. Through the blackness Mark felt his way to the car. In those days he kept his keys in the car, behind the visor. But in the dark his hand hit the visor, and the keys fell to the floor. The next minutes, which seemed like hours, were spent scrambling with shaking fingers over the car floor. Above the pounding of his heart, Mark could hear the splintering of wood and rushing of water. At last he found the keys and started his car. Just as he reached the street, however, another driver came along and parked across his driveway. The newcomer left his car, ran across the street to the park, and began to climb a utility pole. Mark ran to the foot of the pole and shouted up to the climber to move his car. "Forget the car!" commanded the voice from above, "The wave is coming now. Get up a pole!" Obediently, Mark began to climb, only to be stopped in his ascent by frenzied shrieks. "Not this one, not this one, go to the other pole!" Too bemused to argue, Mark jumped to the ground, ran to the next pole, and scampered up it just ahead of the rising water. He climbed about three-quarters of the way to the top—the water reached his feet. When it had subsided, Mark and his new acquaintance left their perches and drove off in their respective cars. This proved to be less than a good idea for Mark. As he drove into the Waiākea district the water rose again, causing his car to stall. Hearing calls for help, he left his car and went across to a house to assist a woman who was trying to escape from her upstairs window. The woman landed on top of him when she jumped from the window—immersing them both in the muddy flow. They rose from the flood, however, and were able to make their way on foot to higher ground.

In spite of their terrifying experiences, people like Mark Olds, the Okamotos, and the Shigemasas remained where they knew there was land underneath them. But there were others who would feel the full power of the ocean, as Herbert Nishimoto had at Laupāhoehoe in 1946.

The 1946 waves had done little damage on the Puna (south) side of the Wailoa River. Houses along the river had been flooded on the ground floor, but none had been destroyed and no lives were lost. In fact, on that occasion many residents from the neighborhood had run to the base of the big mango tree near Fusayo Ito's home. Because of this experience, Mrs. Ito and many others decided to stay in their homes. Mrs. Ito's recollection of 1946 was one of excitement rather than terror. She remembered hearing the approach of the waves and thinking it was the noisy baseball team from Honolulu, just disembarked from the boat. So, despite her daughter's pleas that she should leave, on this occasion she opted to stay. Mrs. Ito would have liked to have gone down to the riverbank like most of her neighbors, but she didn't want to venture out in the dark. Instead, she watched from her door until after midnight, when people began to walk by saying that the time of danger had passed and nothing had happened. For a few moments she felt a great relief of tension—then her world was shattered. Her heart leapt as she heard the alarming sound of an explosion "like a bomb," and she was enveloped in darkness. In the next instant the wave entered her open door, seized her and spun her around and around, and churned everything in her home. She was hit on the head, fell through a hole, struggled to lift herself, and lost consciousness.

The next she knew was the sensation of being among bushes. How could bushes be in her house? Maybe her house had been moved to the river's edge. Eyes tightly closed in fear, she became aware of the sound of water. She moved one leg and tried to find a foothold—but met nothing solid, only water. For a while she floated on her back, then slowly opened her eyes. Above her in the great black expanse shone the stars. "Then that's the first time I cried—cry, cry, cry, because I was so, so scared—nobody around." Then she was deluged by another wave: water ran up her nostrils, then in the thick darkness she swallowed more water. The smell of the gasoline on the surface of the water was terrible. She was whirled around and propelled toward the bay. Borne up by a piece of debris, she heard whistling nearby and a man's voice calling, "Can you swim?" "No," she replied. "Then hang on!" Grasping the piece of debris, she was dragged by the ebbing water past the Hilo Ironworks and between the tops of two large pine trees. Moving very fast, Mrs. Ito was washed down by Wainaku mill, saw its light, then found herself in the ocean.

Her eyes had now adjusted to the darkness, and she could see the vast amounts of debris that surrounded her. In her confused state she thought the surface was so thickly covered with wreckage that she would be able to walk back to the shore. She put one leg over the edge of her makeshift raft and realized that the water was very deep. She could see the lights of vehicles moving along the shore but knew she could not reach them. All alone on the ocean, she heard no sound but the roaring of breakers. About this time, she became aware that she was being supported in the ocean by a window screen from her house. Only a tiny woman like herself—a mere 4 feet, 11 inches tall—could have been kept afloat by such a flimsy structure. All night she was tossed on the turbulent water, lifted on the crests of the waves, then plunged precipitously into the troughs. By this time no land was visible, but she saw some lights far away. Then there was only "sky and ocean, sky and ocean." Mrs. Ito cried for a while, then made peace with her God. The sky was beautiful. She accepted that death would come eventually, by sharks or by drowning, but she felt no concern. She had no control over whatever might happen.

By 2:15 A.M. the height of the waves reaching the bayfront had diminished, and the geologists from the Volcano Observatory felt it would be safe to enter downtown Hilo to assess the damage. They were alarmed at what they found.

> Thick slimy mud covered the streets, and fish abandoned by the water that carried them over the sea wall were strewn about. Hilo's sewage, dumped inside the harbor entrance, had been stirred up by the first two waves and hurled into the face of the city by the third, filling the air with a distressing stench. At the north end of Kamehameha Avenue damage was slight, consisting only of broken windows and muddied floors. Stores in the block north of Haili Street had been breached by the waves, which gathered up their contents and dumped them in confusion on the street. Broken power poles, tangled wires, yardage goods festooned through wires and muck, children's toys and gasping fish clogged the street, making our progress southward treacherous and slow.

As the geologists continued to work their way through the devastated town, they saw four people emerge from a second-story window in one of the few buildings left standing. The stairs had been carried away as the waves had gutted the ground floor. The geologists helped the survivors down to the street. "Only then, as we picked our way northward along Kamehameha with our unexpected charges, did the horrible reality sink home: Hilo's streets had been evacuated, but its buildings had not!"

Rescue operations continued throughout the night and into the early-morning hours. Once again the fire department and the police department were very busy, along with countless volunteers. Neighbors and friends helped each other where they could, but it was difficult to make much progress in the deep darkness that followed the loss of electricity.

The police had been put on alert, but when midnight arrived and nothing happened, and they heard reports of low wave heights in Kaua'i, many officers, including Bob "Steamy" Chow, thought that this was another false alarm. His detail had been to evacuate downtown Hilo, but when everyone thought the danger was over, people started to go back to their homes—there was insufficient manpower to stop them. At about 1 A.M. Bob Chow had been on Pu'ue'o Bridge. When the water went out of the Wailuku River he decided to move, and he was just getting into his car when the wave hit the first road bridge. When he reached Shipman

Street he saw the flash light the sky as the HELCO plant was hit, then the city went dark. Along Kīlauea Avenue the road was wet, but there was no sign of damage until he reached Furneaux Street. There he heard screaming; a man was terrified on the second floor of the Hualani Hotel—but he was unhurt. Mamo and Keawe streets were blocked by rocks, so on foot he set about the task of keeping people out of the stricken area. On the south side of Mamo Street people were trapped in their houses and calling out, but help was at hand in the form of the County Rescue crew, with chain saws. The weather was fine, and the rescuers were not hindered by rain. Steamy Chow stayed at his post all the next day. His wife, Lily, did not know what had happened to him; she had heard rumors that a police officer in Waiākea had been killed. In fact that officer, Godfrey Desha, escaped with his life by running from the wave, leaving his car behind. The car was washed under a building and sliced in half so that it looked like a convertible.

All through the city people were trying to salvage possessions in their homes and businesses. Martha van Gieson's home in Waiākea, where she lived with her fiancé's sisters, was washed away, its remains floating in Reeds Bay. Her Bible alone was preserved, found floating on a mattress.

Thirteen-year-old Thor Wold had gone with his father to see what had happened to their Hawaiian Fernwood Lumber Mill, located across the road from the Civic Auditorium. It was still there, but the floor of the warehouse was deep in mud and sewage, releasing a terrible stench Thor remembers to this day. Stuck in the mud at the back of the warehouse were two boats that had been swept there by the wave.

At Koehnen's Furniture Store on Kamehameha Avenue, Carl Rohner had been working all night with his wife, Helie (Koehnen), and daughter Karyl. Carl had heard on the radio that there was water in the Hilo Theater, so he set off downtown. On his way along Kīlauea Avenue he heard the crash of houses being torn from their foundations, then saw the flash as the wave hit the HELCO plant. He kept going, persuading the police that he should be allowed through, but of course when he arrived at the store he found it in complete darkness. He returned home for lights and assistance, and went down to the store with Helie and Karyl to push out water with one-by-three boards taken from the bed frames in the store. Their banker, Chuck Vanatta, arrived, still dressed in his suit and tie from a party, and got right down to help them.

Meanwhile the rescue efforts continued, as there remained a major concern about the number of people missing, especially given the possibility that many had been washed out to sea by the giant waves. Indeed, there were such victims: as Monday, May 23, dawned in Hilo, Mrs. Ito was still alive, still floating offshore.

The morning tide carried her back toward Hilo. As the light spread across the water, she saw that most of the debris had dispersed during the night and she was alone on her window screen. Eventually she saw something white, and wondered if it could be a "ghost ship," a figment of her imagination. In fact, it was the Coast Guard's 95-foot patrol boat, under the command of Chief Boatswain Fredrick R. Nickerson. Mrs. Ito had been spotted about 800 yards from the boat by James Alexander as he scanned the debris through his binoculars. Almost at once one of the sailors, John Harris, at his lookout on the mast, saw someone waving. As the boat approached the lone figure, Harris and another sailor, Thomas Williams, jumped into the water and swam to her assistance. When Fusayo Ito heard the splashes and saw her rescuers, all the peace and resignation she had felt during the night left her. First she was rigid; then she surrendered herself to their ministrations. She was lifted aboard, given first aid, and wrapped in a blanket. After her long ordeal her only physical injuries were a cut finger and bumped knee. But the effects of the shock were great, and for many weeks the sound of water would set her shaking.

After striking Hilo, the waves had continued in their path across the Pacific. In Oregon, according to reports in the *Seaside Signal* on May 26, the tsunami "brought surges of high water in the Necanicum River over a period of 48 hours." One bore was described as being almost 5 feet in height, damaging boat landings, swamping boats and drowning one unidentified man. The largest bore "shot up the river" at about 9 A.M., on the morning of May 23. In Crescent City, California, three commercial fishing boats were sunk, and in Los Angeles 40 boats were sunk and another 200 damaged by the surges. There was also one casualty in southern California. A skin diver searching for abalone during the tsunami was reported missing and presumed drowned by the waves.

On the opposite side of the Pacific, along the coast of the Kamchatka Peninsula and in the Commander Islands of the Russian Far East, tsunami waves between 18 and 22 feet were recorded.

Even at the entrance to the distant Arctic Ocean off Alaska the effects of the tsunami may have been felt. In the early afternoon of May 23, Eskimos out on the ice of the Chukchi Sea near Point Hope heard ice cracking and quickly returned to shore. The tsunami would have had to pass through the Bering Straits and travel under the still-frozen oceanic pack ice. Then, as the waves moved into shallow water and increased in size, they became large enough to crack the thinner coastal ice.

But it was perhaps in Japan, despite all its experience with tsunamis, that the disaster was least expected. Just before dawn on May 24, a full day after the earthquake and 10,000 miles away from Chile, waves as high as 25 feet began to come ashore along the Japanese coast. Property damage of more than $50,000,000 occurred in the Tohoku and Hokkaido districts, and the death toll reached 140.

Although the 1960 Chilean tsunami caused only moderate damage on

Figure 5.12 Tsunami waves from the earthquake in Chile running upstream at the mouth of the Niita River, Hachinohe, Aomori Prefecture, Japan, at 9:30 a.m. on May 26, 1960.

most of the Hawaiian Islands, in Hilo it had been catastrophic. The first wave had reached Hilo at 12:07 A.M. local time, and, as in 1946, 1952, and 1957, the first indication was a rise in the water level. The tsunami traveled the 6,600 miles from Chile in 14 hours and 56 minutes—an average speed of 442 miles per hour. The arrival time predicted by the Tsunami Warning Center had been 20 minutes earlier than the actual arrival time in Hilo, an error margin of only 2 percent.

Since the two tide gauges in Hilo Harbor were put out of action by the waves, the best record of the movement of the water in Hilo Bay during the tsunami was provided by the observations of the geologists from the Hawaiian Volcano Observatory. They recorded a period between waves of just over half an hour for the first two waves, but the third wave was very different. Ten minutes before the third crest should have washed into Hilo, a giant vertical wall of water—a bore—advanced on the town.

The popular belief that a great withdrawal of water from the shore precedes a giant tsunami wave may be more fact than fiction. A tsunami wave may be transformed into a bore when it advances at sufficiently high speed through very shallow water. The formation of the bore in Hilo Bay may have been initiated by the large withdrawal of water during the trough of the second wave. Descriptions from observers near the Wailuku River and reports from boatmen near the harbor entrance indicate that before the bore arrived, the water in the harbor was about 7 feet below its normal level; the bore formed initially at the harbor entrance, near the end of the breakwater, where the depth of the water becomes shallower. Evidence suggests that the height and turbulence of the bore reached a maximum in the bay between the harbor entrance and the Hilo bayfront, like enormous breaking surf. At this point, it was estimated to have been traveling at more than 30 miles per hour. The wave height along the shore steadily increased southeastward along the bayfront, reaching a towering 35 feet near the Wailoa Motors building, located at that time on Kamehameha Avenue. The final flood of water into Hilo was described by scientists as being "analogous to the sheet of water that races up a beach beyond the spent breaker that propelled it."

As the wave surged into Hilo, it wrenched 22-ton boulders from the 10-foot-high bayfront seawall and carried them as far as 600 feet inland across Moʻoheau Park, without leaving a noticeable mark on the lawn. The water struck with such force that 2-inch pipes supporting parking meters along the waterfront were bent over parallel to the ground. Electric cables and transformers were torn from utility poles. The reinforced concrete office of the Hilo Iron Works withstood the force of the wave, but its second-story skylights were blasted out by the increase in air pressure as the wave struck the building—it had become a manmade "blowhole." At a nearby showroom, an 11-ton tractor and the building housing it were removed by the wave. Bobby Fujimoto's Hawai'i Planing Mill lay directly inland of the maximum 35-foot wave height measured along

Figure 5.13 Hilo bayfront parking meters bent parallel to the ground by the tremendous force of the waves of the 1960 tsunami.

Figure 5.14 Aerial photograph of Hilo showing 1960 tsunami destruction in the central business district.

Kamehameha Avenue. The large, modern hardware store, made of prefabricated steel, was completely swept away, right down to its tiled cement floor. The only remnants were ½-inch steel reinforcing rods sticking up from the foundation. Pieces of the building were later found 1,500 feet away beyond the Wailoa River.

In this zone of total destruction, entire city blocks were swept clean. Buildings were wrenched from their foundations and deposited as piles of debris hundreds of feet away. Some light frame structures were floated off their foundations and deposited elsewhere, sometimes in clusters, without major damage except by collision with other structures. Others were pulled into Hilo Bay, some drifting offshore and breaking apart—littering Hawai‘i's shores for weeks afterward. Scores of automobiles and trucks were totally wrecked. In some cases, cars were stacked three-deep by the waves. Nearly all of the fishing fleet that had not put out to sea was destroyed by the waves. Thirty-foot lengths of concrete curbing from the Bayfront Highway were carried 350 feet inland. And once again, the strong footbridge to Coconut Island, replaced after the 1946, 1952, and 1957 tsunamis, was swept away. All of this damage was done by the third wave.

Figure 5.15 Mail truck smashed into a house in Hilo, part of the damage caused by the tsunami of 1960.

Figure 5.16 Fishing sampan washed ashore and overturned by the force of the waves at Hilo in 1960.

Even to this day, scientists are not completely sure exactly how the bore formed and arrived 10 minutes before the third crest. One suggestion was that a harbor seiche—the natural surge of water in and out of the bay—had acted in concert with the advancing crest of the third wave. But the arrival of the bore 10 minutes before the third crest argues against this hypothesis. It was more likely that the harbor seiche added to the withdrawal of water during the second trough, which then lowered the water level and set the stage for the bore to form. In any event, after the third wave the period between waves was 15 minutes rather than 34 minutes, the same periodicity as occurred in the 1952 and 1957 tsunamis—probably representing the natural period of oscillation of Hilo Bay.

On the other Hawaiian islands the tsunami reached moderate heights and caused moderate but widespread property damage. With the exception of Hilo Bay, wave heights on the island of Hawai'i itself rarely exceeded those of the 1957 Aleutian tsunami, and averaged only 9 feet. There were some local variations in wave heights, mostly in V-shaped inlets or shallow bays, with 17-foot waves recorded at Honu'apo and Ka'alu'alu, and 12-foot waves at Honomū.

Along the west coast of the island a wave height of 12 feet was reported in the small bay at Keauhou, and a height of 10 feet reported at Kahalu'u. Only at Nāpō'opo'o were the waves of the 1960 tsunami larger than those of 1946 or 1957: waves 16 feet high washed over the settlement, destroying six houses and moving a number of others off their foundations. Eyewitnesses told of waves coming from the north and washing southward over the low-lying village. The movement of buildings and debris also indicated that the waves came from the north, not from the south—the direction of Chile. How could this be the case? Nāpō'opo'o is situated on the south shore of Kealakekua Bay and lies opposite the steep cliffs of the Kealakekua fault zone, which form the north shore. The destructive waves were, perhaps, the result of the main tsunami waves (coming from the south) bending into the bay and being reinforced by the reflected tsunami waves bounced off the cliffs. This combination produced the 16-foot waves that washed over the village.

The tremendous damage done by the waves in Japan was more difficult to explain. Scientists now believe that there were three main reasons why the tsunami was so large and destructive there. As in Hilo Bay, resonance between the tsunami waves and the period of seiche of bays along the Sanriku coast may have played an important role in increasing the size of the waves at the heads of these bays.

A second reason for large waves impacting the coast of Japan has to do with the geographic locations of Chile and Japan, which are at opposite ends of the Pacific, and the shape of the Pacific Ocean basin in between. Imagine a ripple spreading out from the edge of a pond. As it heads toward the middle, the ripple crest is stretched completely across the pond. Then as the ripple approaches the opposite side, the length of the crest is compressed into a shorter and shorter distance. As the length of the crest is stretched, its height decreases; conversely, as the length is compressed, the height increases. As the tsunami waves spread out from Chile across the Pacific, they were forced to cover a larger and larger distance until they had reached the middle of the Pacific Ocean basin. But then as they approached Japan, the length of the tsunami crests would cover a progressively shorter distance, resulting in increased wave heights along the coast of Japan.

This being the case, why should there have been such large waves in Hawai'i, lying as the islands do in the middle of the Pacific Ocean? This brings us to the third reason the waves were especially large in Japan. The depth of the Pacific Ocean basin varies enormously from deep abyssal hills and plains at around 18,000 feet to the shallow, young ocean spreading centers like the East Pacific Rise at depths of only around 8,000 feet. Because the speed of tsunami waves is controlled by water depth, the parts of a tsunami crest that pass over the East Pacific Rise or other shallow features such as the Hawaiian Ridge or the Ontong-Java Plateau slow down. Meanwhile the parts of a tsunami crest on either side, still in deeper water, continue at a greater speed. This results in the tsunami wave crest bending toward and around shallow features. The process is called wave "refraction" and can produce a focusing of wave energy, much like light waves can be focused with a magnifying glass to burn a hole in a piece of paper. The shallow features of the Pacific Ocean floor are arranged in such a way that a tsunami generated in Chile will be strongly focused along the coast of Japan and to a lesser extent on the Hawaiian Islands. The combined effects of resonance in coastal bays, geographic compression, and focusing of wave energy had resulted in tsunami waves from Chile causing devastation along the coast of distant Japan, nearly half a world away.

Damage in Hilo

As in 1946, the city of Hilo suffered the most extensive damage in all of the Hawaiian Islands. The business district along Kamehameha Avenue and the adjoining low-lying residential areas of Waiākea and Shinmachi were literally wiped off the map. Damage to property included 229 dwellings and 508 businesses and public buildings. The floodwaters inundated approximately 580 acres between the Wailuku River and the shoreward end of the breakwater. Between the Wailoa and Wailuku rivers the water washed inland as far as the 20-foot contour above sea level. Property damage was estimated as high as $50 million.

But the greatest cost could not be measured in dollars. In Hilo, 61 people were crushed or drowned by the tsunami waves, and another 55 people were injured, requiring hospitalization and medical care. Because all medical facilities had been put on alert, they were ready to receive the casualties. Some injuries were slight, like the cut under Robin Spencer's eye. About 75 such injuries were treated at the outpatient department of Hilo Memorial Hospital, and another 350 at the first aid station at Hilo Intermediate School. Taffy Okamura had gone there, after the water receded from his lauhala tree refuge. He and his friends had moved through the darkness, the water, and the debris, tripping over fallen cables. At the civic building they met members of the fire department and were taken to the school. After some warming coffee, Taffy helped prepare beds for the victims, went home to his family at 4 A.M., and was back to work at daylight to help with the cleanup.

More serious cases of injury were admitted to the care of Dr. Taniguchi and his colleagues at Hilo Hospital. All these casualties had the kind of injuries typical of rough battering by debris-filled tsunami waves. In addition to this, survivors soon began to show signs of having been exposed to severely polluted water. Two patients, both women, died within 48 hours; all the others survived, but some endured long illnesses. Some had been peacefully asleep when the wave hit, like a twenty-five-year-old man, who had no recollection of what had happened to him or how he sustained the severe injury to his left foot. The wound became so severely infected that the foot had to be amputated. A seven-months-pregnant Filipino woman was admitted with a broken leg, rib fractures, and fractures of the pelvis. Miraculously she delivered her healthy baby on August 20 without any complications.

Others, like a seventeen-year-old Hawaiian boy and a forty-seven-year-old tourist from California, paid a high price for the rash decision to watch the waves rather than seek safety. The boy had been on foot, and he was carried hundreds of feet inland by the wave. Despite his multiple cuts and bruises, he had no broken bones. However, the enormous force that caught him had ruptured his esophagus. Ingesting the sewage-laden waters of Hilo Bay led to a severe infection and fever. The infection was difficult to combat, and he was fortunate that a wide range of antibiotics was available. His treatment was completely successful—but he did not leave the hospital until August 1.

The Californian tourist had been at the bayfront with some friends. Like many of the onlookers who had been on the banks of the Wailoa River with Al Inoue, she had her car close by, planning a quick escape if necessary. But when the wave loomed before her and she jumped into the car, the engine would not start. The car was swept inland about 400 feet. Water entered her stomach and her lungs, causing pneumonia and severe infection. As with the rash teenager, the multiple antibiotics and the care she received in the hospital saved her from being added to the roll of the dead, and she returned to California on July 26. The multiplicity of bacterial organisms present in the tsunami floods, and the crucial role of antibiotics in combined doses, was noted by Dr. Taniguchi in a paper he published with Dr. Woo in 1961.

Fortunately, no lives were lost elsewhere in the Hawaiian Islands. But once again, Hilo had been brought to its knees. Until the water mains were repaired, residents were urged to boil their water. The pumps of the local sewage system had been destroyed, leaving many areas with no sewer service. The destruction of the electric power plant left the electric company unable to meet the power needs of the island. The County Health Department advised that damaged food stocks be inspected before use. Milk had to be brought in from Honolulu because the local pasteurization facilities had been destroyed. Fog machines were brought in to spray insecticide in potential mosquito-breeding areas. Public shelters had to be used to house 215 families until other arrangements could be made. Many people turned out to help those who had lost their homes and businesses. They took rolls of fabric from the dry goods stores to Carvalho Park, washed them in the stream, then laid them on the ball park to dry like brightly colored patchwork quilts in the sun.

The devastated areas had to be cleared—no one knew what would be found under the wreckage. Equipment and manpower were required. Both existed on the sugar plantations. The Corps of Engineers contacted Yasuki Arakaki and asked him to organize the clearing of the area from Waiākea town down to the Hongwanji. He used a bulldozer with its push rake from ʻŌlaʻa Sugar, and got plenty of volunteers from the cane workers. They worked on the cleanup after they completed their 8-hour shift at the plantation, sifting the debris carefully in case there were any bodies and then loading it into trucks loaned by Yamada Trucking. All the workers volunteered for the task, but exactly a year later the Corps of Engineers contacted Yasuki to find out how many hours each man had worked so they could receive payment. This was an unexpected and pleasant reward for what had been a hard and unpleasant task.

There were some pleasant surprises in the aftermath. Like many other Hilo residents, Mrs. Ito had lost her home in the deluge of the Chilean tsunami. She had no recompense and asked for none, thankful to have survived. In her hospital bed, she had wept tears of gratitude, "happy, happy, happy" to be alive. She continued to feel that way even though her life savings, in the form of savings bonds, had been swept away with her home. But soon she was to be even happier, and her astonishing story was to have a surprise ending. All her important papers, including the bonds, were kept in a waterproof bag. When clearing work was being done on the side of the river opposite the site where her house had stood, the bulldozer driver was stopped by an obstruction. When he went to clear it he found a waterproof package, which he took to the police station. What a wonderful surprise when the police called Mrs. Ito, who recounts: "I cried and cried—if I was dead, I don't need those things, but if I am alive, I need them." Earlier, she had tried to claim the value of the bonds from the bank but had not known the serial numbers. It took her days to soak the mud from the bonds, but in the end the serial numbers were visible and the bonds were honored by the bank. Fusayo Ito felt that her Buddha was looking after her.

What Went Wrong?

The 1960 Chilean tsunami, like that from the Aleutians in 1946, had been a major catastrophe for Hilo. In the wake of the destruction the authorities began to count the cost of the disaster and to try and determine why there had been 61 deaths. It was time to figure out what had gone wrong with the system and what could be done to prevent another tragedy in the future. It was also time to rebuild.

Kaiko'o

Following the 1946 tsunami, the strip of land between Kamehameha Avenue and the bay front had been converted into a recreation and parking area that was to serve as a buffer zone against future tsunamis. Many of the businesses displaced from this buffer zone had been rebuilt in other low areas. Some of the residential communities that had been severely damaged in 1946 were rebuilt with crowded, flimsy houses. The impact of the Chilean tsunami waves on these residential and business areas, particularly at Waiākea and Shinmachi, was the cause of much of the damage and loss of life in 1960.

It was realized at last that Hilo would always be at the mercy of destructive tsunamis. Just 8 days after the 1960 tsunami, the Hawai'i Redevelopment Agency was established. The ocean side buffer zone was extended and a landfill plateau was constructed, raising the inland border of the greenbelt 26 feet above sea level. The project was named Kaiko'o, or "rough seas," and it would soon become the new commercial center of Hilo. Federal and state funds for public housing and urban renewal were provided, and the Small Business Administration made loans available to help businesses get started again. Because the local population was justifiably skeptical about rebuilding, even behind the buffer zone, the state and county buildings were the first constructed. They now stand on the bluff overlooking the buffer zone, just seaward of the Kaiko'o mall.

The Warning System

The Tsunami Warning System was, perhaps, the greatest success and at the same time the greatest failure of the 1960 catastrophe. The system had warned of the approaching waves and accurately predicted their arrival times, but it had been apparent that public education as to the nature and seriousness of the tsunami had been totally ineffective.

During the 1952 and 1957 tsunamis, large numbers of people had failed to leave danger areas when told to do so. Instead, sightseers had converged on the coast. Thanks to the small size of the waves, there had been no loss of life. But the behavior of the sightseers should have served as a warning that the next large tsunami would kill many people. After the 1960 tsunami, newspaper headlines told of Civil Defense officials being shocked by the reactions of residents; however, people had reacted no differently than they did in 1952 and 1957.

Many people returned to danger areas after the first small waves had passed and were then overwhelmed by the giant bore. At Kūhiō Beach on O'ahu, people who should have known better ran out on the reef and picked up fish. Police set up roadblocks, but they were unable to prevent sightseers from entering dangerous beach areas. At the Moana Hotel in Waikīkī, guests refused to leave the Kamaaina Bar despite warnings from the management. The guests were lucky and survived, even though the bar was inundated by the waves. State Governor William Quinn pleaded with residents over the radio, but residents and visitors alike flocked to the beaches to witness the tsunami.

A study of the behavior of Hilo residents during the 1960 tsunami revealed that almost all of the 329 adults interviewed had received some kind of warning that a tsunami was impending. Yet, only 32 percent evacuated after they received the warning. More than half simply waited at home for more urgent or specific instructions until it was too late and the waves struck.

Just after the tsunami struck, the governor declared a state of emergency throughout the state, which empowered the National Guard to maintain order and prevent looting. It was debated whether or not the National Guard should have been used before the tsunami struck to enforce orders to keep the public out of danger areas. Oddly enough, it was argued that enforcement might infringe on a citizen's "constitutional right" to endanger himself by remaining in a threatened area.

The most glaring fact is that the public had only vaguely understood how the warning was to be given and how they were expected to respond to it. After 1960, the warning procedure was changed, and the current alarm system was established. This alarm procedure is as follows:

Three hours before the arrival of the first wave, the Civil Defense sirens will sound the "attention/alert signal" (a 3-minute steady siren tone, repeated as necessary). The attention/alert signal means "turn on your radio." Radio stations switch to State Emergency Alert System (EAS) status, and regular announcements are made about procedures to follow and giving current tsunami information. The attention/alert signal is sounded again 2 hours, 1 hour, and 30 minutes before the estimated arrival time of the first tsunami waves. Each of these four signals is accompanied by emergency announcements over the EAS.

The State EAS system and tsunami evacuation areas are described in the front pages of the telephone directory. The first workday of every month at 11:45 A.M., the Civil Defense sirens are sounded as a test. But if the population remains ignorant of what the sirens mean or fails to respond to the warning, the next destructive tsunami could well be another killer.

Tsunami Research

The 1960 tsunami had given scientists another "live experiment" to study how tsunami waves behave in the Hawaiian Islands. Immediately following the tsunami, researchers set about collecting data. The precise directions of wave advance were determined by measuring what had been wrought by the tsunami's force: the compass bearings of bent parking meters and traffic signs, the direction of gouge marks left by buildings and other heavy debris, and the orientation of lines between sites where buildings had been deposited by the waves and their locations prior to the tsunami.

The heights reached by the waves were calculated by noting the level reached by the water as indicated by various kinds of evidence: salt-killed grass and other vegetation along the shore; persistent strand lines of fresh cane trash; abraded bark and broken tree branches; debris left hanging on fences, buildings, and trees; and water stains and abrasion on buildings. The distance of each of these features above sea level was carefully measured.

The research results show a striking contrast between the 1960 Chilean tsunami and the 1946 Aleutian tsunami. The waves from the 1946 Aleutian tsunami reached a maximum of 55 feet and averaged 30 feet along the northeast coast of Hawai'i. This is the coast facing directly into waves from the Aleutians and the heights on this side average more than twice those of the island as a whole. A similar increase in wave height might have been expected along the southeast coast in 1960; this side of the island most directly faces Chile. Waves were no larger there, however, than along the more protected west and northeast coasts. Specialists believe that these variations in wave pattern might be caused by the difference in cross-section of the Hawaiian Ridge as encountered by waves approaching from different directions. The ridge presents a barrier of almost continental dimensions to Aleutian tsunamis, which approach it broadside, but presents only a small barrier to Chilean tsunamis, which approach it end-on. Another factor that might have led to larger waves along the northeast coast is the relatively shallow, sloping shelf that extends outward from the sea-cliffed Hāmākua coast north of Hilo and is absent elsewhere on the island.

The 1946 and 1960 tsunamis also show a marked difference in the pattern of flooding in the Hilo area. In the central part of downtown Hilo (northeast of Kumu Street), and east of the Waiākea Peninsula, the 1946 waves flooded inland farther than the 1960 waves. The 1946 waves had been relatively high (17 to 25 feet) throughout the entire bayfront area from the Wailuku Bridge to the Waiākea Peninsula, but were lower in the lee of the peninsula. In 1960, the waves were large even in the lee of the peninsula. In fact, between downtown Hilo and Reeds Bay, flooding was almost twice as extensive in 1960 as in 1946.

From their studies the scientists concluded that though each tsunami is indeed unique, the location of a tsunami's source may be one of the major factors in determining which areas will be most affected by the waves, and how great the subsequent damage will be. Tsunamis from near the same geographic place of origin tend to produce remarkably

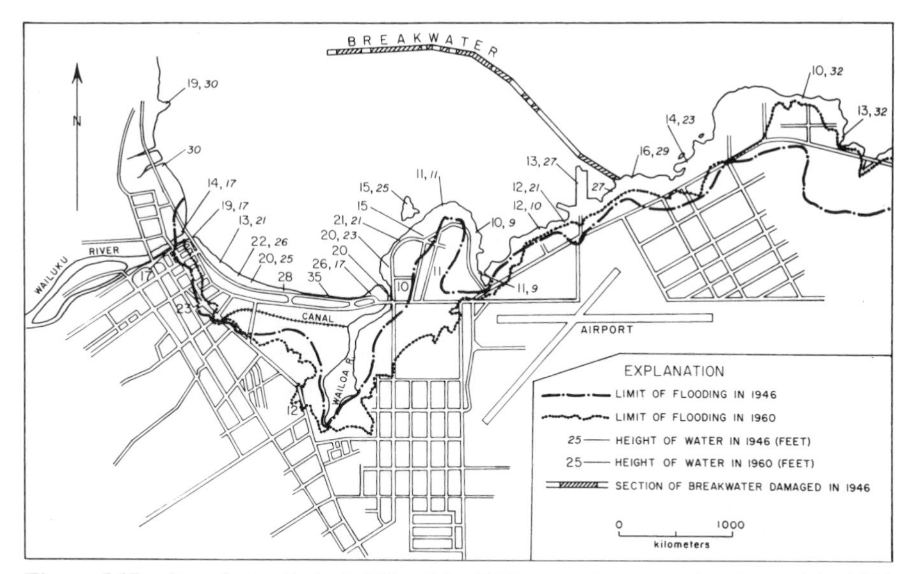

Figure 5.17 Inundation limits in Hilo of the 1960 tsuṇami compared to those of the 1946 tsunami.

similar relative patterns of wave height at any one location. For example, the tsunamis of 1946 and 1957 both originated in the Aleutian area, and though the maximum wave heights in Hilo from the two tsunamis were very different, both severely affected the same parts of town. The 1960 Chilean tsunami, on the other hand, produced a very different pattern of inundation from either of the Aleutian tsunamis. In other words, the patterns of wave heights and hence flooding on Hawai'i's shores produced by tsunamis of different geographic origin are strikingly different. Tsunamis from nearly the same place of origin, though perhaps differing in relative severity from place to place, tend to be more similar than scientists previously thought.

The scientists also learned more about how tsunami waves interact with coral reefs and with small, steep islands. The radio announcements in 1960 of the small wave heights at Tahiti gave a false sense of security to many in Hawai'i. Tahiti is surrounded by coral reefs. Reefs tend to break up the tsunami waves, dispersing and absorbing their energy. As one scientist put it, "Tsunamis can go past day after day and they'll hardly even know it."

Small, steep islands also tend to be minimally affected by tsunami waves. One of the Line Islands, Kiritimati, located about a thousand miles south of Hawai'i and previously called Christmas Island, is one of the very few tide stations lying between the South American coast and the Hawaiian Islands and could have served as a valuable indicator of the severity of a tsunami heading toward Hawai'i. The 1960 Chilean tsunami produced a 6-inch rise on the tidal record at Christmas Island. In 1960, this small rise was interpreted to mean that there was a tsunami, but a small one. As one of the scientists said, "We later found out that a 6 inch rise at Christmas Island is a big tsunami! These are the things that you're only going to learn after they happen. This is how we refine the art, so to speak. The more tsunamis we have, the better we'll get." Scientists at the warning center had learned painful lessons from the 1960 tsunami.

So, too, had the inhabitants of other Pacific islands. At Pitcairn Island, in New Guinea and New Zealand, and in the Philippines and on Okinawa, some 300 lives were lost.

The great destruction caused by the May 1960 Chilean tsunami prompted Japan and a large number of other countries and territories to become members of the Tsunami Warning System.

In Hawai'i, the Hilo Bay area had proved to be the most vulnerable area of the islands. The Chilean tsunami had been bent, bounced, and funneled into the bay with disastrous results. The 1952 Kamchatka tsunami had radiated around the island and reached its maximum heights in Hilo Bay. Hilo Bay faces directly toward the Aleutians, source of the 1946 and 1957 tsunamis. What would a truly large earthquake in Alaska, bigger even than the one in 1946, do to Hilo?

6

The 1964 Good Friday Earthquake and Tsunami

Two giant tectonic plates collide along a boundary made up of the Aleutian Islands and the Pacific coast of Alaska. Here the oceanic Pacific plate thrusts under the continental North American plate, forming what is known as a "subduction zone." The Aleutian-Alaska Trench is the surface expression of this subduction zone, and ideal conditions exist here for the generation of tsunamis—that is, large earthquakes associated with vertical motions of the sea floor.

The tsunamis produced along the Aleutian Trench, such as those of 1946 and 1957, are a threat mainly to the Hawaiian Islands. Tsunamis generated in the Gulf of Alaska, on the other hand, pose a threat to the population centers of Alaska itself as well as to the west coast of Canada and the United States, due to the orientation of the tsunami-generating faults.

At 5:36 P.M. Alaska Standard Time on Good Friday, March 27, 1964, one of the largest earthquakes ever recorded in North America struck Alaska. The quake was centered near the eastern shore of Unakwik Inlet in northern Prince William Sound, a sparsely settled region of high, rugged topography and numerous glaciers. When the magnitude of the tremors was later determined, they would measure an awesome 8.4 on the Richter scale, releasing twice as much energy as the famous 1906 earthquake that destroyed much of San Francisco.

The earthquake was produced by movement along a complex fault dipping from near the Aleutian Trench beneath the continent along a hinge line running roughly northeast-southwest parallel to the southeast coast of Kodiak Island; it had its focus at the relatively shallow depth of about 14 miles. The quake was accompanied by vertical movement over an area of as much as 200,000 square miles, which ranged from subsidence of 7½ feet to uplift of as great as 38 feet and included the area of the islands and mainland of Prince William Sound.

The nearest populated areas to the epicenter were the small communities of Valdez, 40 miles to the east, and Cordova, 70 miles southeast, and the city of Anchorage, 80 miles to the west.

As the seismic waves radiated from their source, many Alaskans would describe feeling seasick from the motion of the earth. In Anchorage, cars, trucks, and even aircraft were observed bouncing as though they were "on a trampoline." At Valdez the shaking was reported to have lasted from 3 to 5 minutes and to have been accompanied by a "low pitched rumbling sound." The ground was said to have had a rapid rolling motion, "heaving in much the same manner as a ground swell in the open ocean, except that the swells were much more rapid and frequent." Observers estimated these ground "waves" at 3 to 4 feet high. Trees reportedly pitched about wildly as if lashed by storm winds. Large fissures opened in the ground and rapidly filled with water. As the fissures closed, water and suspended sand and silt squirted up from the ground as muddy geysers. After about 2 minutes power poles began going over and

Figure 6.1 Damage in Anchorage, Alaska, after the earthquake of March 27, 1964.

then buildings began to collapse. Concrete and masonry structures suffered severe damage or total destruction, whereas most wooden buildings swayed back and forth with relatively little damage.

These powerful seismic waves set up seiches in rivers, harbors, channels, lakes, and even swimming pools as far away as Puerto Rico and Australia. In the New Orleans, Louisiana, industrial canal, a 6-foot wave tore an 83-foot U.S. Coast Guard vessel loose from its moorings. In Texas, the Coast Guard reported a surge that came through the Sabine Pass and pushed the tide 3 feet higher than normal. At the Tropicana Swimming Club a 2-foot wave splashed over the edge of their pool, causing the loss of 25,000 gallons of water.

The seismic waves generated by the earthquake traveled the 2,400 miles from Alaska to Hawai'i in just 8 minutes, reaching the Honolulu Observatory at 5:44 P.M. (0344 GMT). The observatory staff had gone to their quarters for the evening meal when the seismic alarm sounded. The scientists rushed back to the observatory and immediately noted a large earthquake trace on the seismographs. At 6:13 P.M. they sent out messages requesting seismic data from other observatories. Six minutes later, the first seismic report came in, giving readings from Manila. Within the next 20 minutes reports had come back from Hong Kong, Guam, Japan, California, and Arizona. Yet there were no reports from the Alaskan observatories at College and Sitka. Data from these observatories, as well as tide station data from Kodiak, Sitka, and Unalaska were routed via communications channels passing through the control tower at the Anchorage International Airport. What the scientists at the Honolulu Observatory did not know was that the control tower had been demolished by the earthquake, thus breaking the vital communications link.

By 6:52 P.M., enough information had been reported by various seismic observatories to permit the Honolulu Observatory to locate the epicenter at 61° N, 147½° W, near Prince William Sound, Alaska. At 7:02 P.M. (0502 GMT) the following advisory message was sent out to all agencies in the Tsunami Warning System: "A severe earthquake has occurred. . . . It is not known if a sea wave has been generated. . . . If a wave has been generated its ETA [estimated time of arrival] for the Hawaiian Islands (Honolulu) is 0900Z[1] 28 March."

---

[1]Z, or Zulu, is a short notation for GMT. The notation 0900Z (28 March) is 11 P.M. (27 March) Hawai'i Standard time.

At 7:30, the warning center issued another bulletin, which stated: "Damage to communications to Alaska makes it impossible to contact tide observers. If a wave has been generated the ETA's are. . ."; a list of some 33 sites around Alaska and the Pacific basin followed.

In Alaska, tsunamis were already wreaking havoc. In fact, tsunamis from two different sources struck some areas. In many harbors near the epicenter, the seismic waves caused submarine landslides to occur. These landslides in turn generated highly destructive local tsunamis. Meanwhile, the main movement of the sea floor had generated a major tectonic tsunami that would be felt all around the Pacific basin.

About 20 locally destructive tsunami waves were generated by landslides. These would strike the communities at Valdez, Whittier, and Seward.

Valdez

Only 40 miles from the epicenter of the earthquake lay Port Valdez, the northernmost ice-free port in Alaska, and as such an important terminal for transport via the Richardson Highway into the interior. Situated at the head of Valdez Arm, in the northeastern corner of Prince William Sound, the town of Valdez had been built on the unconsolidated sand and gravel of a delta in the narrow, steep-walled fjord of Port Valdez. Now at the beginning of the heavy shipping season, the town's population had swelled from its official 555 to well over 1,000.

At 4:12 P.M. on the afternoon of March 27, the 400-foot freighter S.S. *Chena*, a 10,815-ton converted liberty ship, had arrived from Seward. Once the *Chena* was tied up at the dock of the freight terminal, nine local longshoremen had gone on board to transfer cargo, and a crowd of 28 adults and children had gathered on the dock to watch. The crew would occasionally throw candy down to the children. Valdez was usually a very quiet place.

It was near low tide when the earthquake struck. According to eyewitnesses, almost immediately the water withdrew from the beaches. With much of the delta on which Valdez was built exposed above sea level, there was a sudden increase in weight on the upper layers of sediment. Ground vibrations from the shock then created a condition in the delta sediments that geologists call "spontaneous liquefaction." The sediments suddenly lost their bearing capacity and a section of the delta, some 4,000 feet long by 600 feet wide and consisting of 98,000,000 cubic yards of material, slid into the sea carrying with it the dock and portions of the town of Valdez.

On board the *Chena*, Captain M. D. Stewart, master of the vessel, was sitting in the dining room when the earthquake struck; he later related: "I made it to the bridge (three decks up) by climbing a vertical ladder. God knows how I got there." The *Chena* then rolled "alarmingly" and pitched violently to port. According to the captain, "The *Chena* raised about 30 feet on an oncoming wave. The whole ship lifted and heeled to port about 50°." Observers on shore said that at this time the ship's propeller was visible. Meanwhile the pier to which the *Chena* was tied was set in turbulent motion. The men, women, and children on the dock ran toward shore or tried desperately to find something to hang onto. Then the dock broke in two, the warehouse flipped forward, and the crewmen of the *Chena* watched in horror as their ship came down on top of the warehouse and dock. In Captain Stewart's words,

> Then it [his ship] was slammed down heavily on the spot where the docks had disintegrated moments before. I saw people running—with no place to run to. It was just ghastly. They were just engulfed by buildings, water, mud, and everything. The *Chena* dropped where the people had been. That is what has kept me awake for days.

Not a single one of the 28 people on the dock survived.

The cannery at Valdez now collapsed into the bay. Captain Stewart knew that he had to get his ship away from shore and out into open water. "I signaled to the engine room for power and got it very rapidly. In about four minutes, I would guess, we were moving appreciably, scraping on and off the mud (bottom) as the waves went up and down. People ashore said they saw us slide sideways off a mat of willow trees (which had been part of the fill material in the harbor)."

Meanwhile, the reflected wave washed back toward the bay. "A big gush of water came off the beach, hit the bow, and swung her about 10° out. If that hadn't happened, we would have stayed there with the bow jammed in a mud bank and provided a new dock for the town of Valdez! The bow pushed through the wreckage of a cannery. We went out into the bay."

The good captain had saved his ship, but two longshoremen, Howard Krieger and Phil Gregordoff, had been killed by shifting cargo in the ship's hold. Another longshoreman, Jack King, was seriously injured and would later lose both feet, and the *Chena's* third officer, Ralph Thompson, died of a massive heart attack.

The first tsunami wave at Valdez had followed almost immediately after the submarine slide and was without a doubt generated by the slide. In Valdez the wave was estimated at from 30 to 40 feet high, but its impact would not be limited to Valdez.

Earlier in the day Red Ferrier and his son, Delbert, had taken their 30-foot boat to a point about 15 miles west of Valdez near the mouth of Valdez Narrows. Needing a tree for timber, they had gone ashore with their skiff when the earthquake struck. The earthquake triggered an avalanche, described as "a 50-foot high snowslide, studded with trees," which roared down gulches on either side of them as they ran for their skiff.

The Ferriers reached their skiff and were only a few yards offshore, frantically heading for their boat about 200 yards away, when the water began to recede. They were left stranded on the sloping beach as they watched their 30-foot boat drop out of sight into a steep channel cut in the rocks. They continued struggling toward their vessel and finally succeeded in pulling themselves aboard, just as they saw a large wave building up at the end of Port Valdez and heading toward the Valdez Narrows. A part of the tsunami wave originating at the port had been propagated westward down the bay. Near Shoup Bay the water surged up over living spruce trees more than 100 feet above sea level, leaving some as large as 2 feet in diameter broken and splintered. At the Cliff Mine, the tsunami ran up to more than 170 feet above sea level and then surged out through the narrows.

The Ferriers raised the anchor, started the engine, and turned the boat south toward open water. Red gunned the engine to full throttle to try and outrun the rapidly approaching wave. As they looked back, the wave soared over the top of the navigation beacon in the center of the narrows, rising perhaps 50 feet high. The Ferriers prayed and continued to gun the engine. They had just managed to clear the mouth of the narrows when the wave overtook them. Fortunately, once outside the narrows the tsunami began to spread out and decrease in height, enabling the boat to ride up over the top of the wave. They described the wave as black with mud, rocks, and other debris. Other large waves would surge past them during the night, but it was not their night to die.

No one in Port Valdez could see what was going on past the dock out in the bay. The turbulence of the first two waves had apparently created a "mist or haze" which obscured their view beyond the shoreline. They couldn't see that 68 of the 70 boats in the harbor had been almost immediately destroyed. Fortunately no one was on board any of them at the time.

But now the sky would be lit up as the tanks at the Union Oil Company, ruptured by either the earthquake or the tsunami, leaked fuel, which caught fire. By 10:30 P.M. the entire waterfront area of Valdez was in flames. Smaller tsunami waves from the main tectonic tsunami washed ashore at 11:45 P.M. and 1:45 A.M., but they failed to put out the fires, which would burn for another two weeks.

Ironically, the landslides that produced the devastating tsunami waves at Valdez could have been predicted. As in Newfoundland in 1929, the submarine landslide that produced the tsunami had also produced a turbidity current, which broke submarine telegraph cables. But this was not the first such turbidity current to break submarine cables off Valdez. An earthquake in 1908 had produced a turbidity current that had broken the cable from Valdez to Sitka, and another earthquake in 1911 had resulted in the burial of 1,650 feet of submarine telegraph cable. There had even been previous submarine landslides affecting the Valdez waterfront. In the early 1920s, a slide during the night had carried off a section of dock onto which a heavy spool of wire had been loaded the day before. In the late 1920s a second slide occurred, and sometime between 1942 and 1945 a third slide occurred beneath the cannery dock, during which the dock was badly damaged though not carried away. But no lives had been lost in any of these slides and little notice was taken. However, following the 1964 earthquake and tsunami and the horrific loss of life at Valdez, the decision was made to move the entire community to a safer area—one that was more stable and had some natural protection from tsunami waves.

Whittier

About 40 miles southwest of the epicenter of the earthquake lay Whittier. Built by the U.S. Army in a long, narrow bay off Prince William Sound, it had once been a large railroad and harbor base. Luckily, by 1964 only 70 people were living there. Almost immediately after the earthquake three waves struck in quick succession. The first wave was low and not particularly destructive, but the second and third waves were 30 to 40 feet in height.

Jerry Ware, his wife, and their six-month-old baby were in their trailer home when the largest wave rushed in. It picked up their trailer along with a railroad car, carrying them both 400 feet up the beach. The trailer was completely smashed and the mother and baby washed clear, but the baby was torn from his mother's arms by the force of the water. Miraculously the baby was later found alive in a snow bank, but he did not survive. In all, 13 people were killed by the tsunami at Whittier. They had all lived close to the waterfront and no warning had been possible. The railroad depot and two sawmills were destroyed. And as at Valdez, fire had broken out at the Union Oil Company tank farm, adding to the destruction caused by the tsunami. A graphic illustration of the force of the water was provided when the waves struck the Two Brothers Lumber Company. The company's 2,300-pound mill was moved about 100 feet, and a 2- by 6-inch plank was driven completely through a 10-ply forklift tire.

Homer

On Homer Spit, a bar extending into Kachemak Bay, the small boat harbor disappeared into a "funnel-shaped pool" within 2 minutes of the start of the earthquake. Then, in not more than a minute, a 26-foot wave washed over the base of the spit on which the town was located and surged into the settlement, covering the floor of the Land's End Hotel, the new Porpoise Room Restaurant, and the Salty Dawg Saloon. This wave was too soon to have come from the major tectonic tsunami and therefore must have originated from locally generated landslide tsunamis. Moreover, the tsunami waves washed in from two different directions, from both the Cook Inlet side and from Kachemak Bay.

Seward

On the southeast coast of the Kenai Peninsula at the head of Resurrection Bay sits the city of Seward. A town of about 1,700 people in 1964, it was

Figure 6.2 A 2- by 6-inch plank driven completely through a 10-ply forklift tire by the force of the tsunami waves at Whittier, Alaska.

an important marine and railroad terminal. Seward patrolman Ed Endresen was in his patrol car when the first tremors struck. A falling chimney landed on the car, damaging it, and he immediately abandoned the vehicle and began running toward the small boat harbor. As he was running, a 30- to 40-foot deep fissure opened up under his feet. As he fell in, he somehow managed to grab hold of the edge of the fissure. Just as he began to climb out, the fissure began to close. Water, which had poured into the bottom of the fissure when it opened, now squirted up, helping the patrolman clamber out. At the last second the fissure closed on his foot. Endresen pulled at his foot, finally twisting it free, but wrenched his back in the process. He then managed to hobble off to safety.

Within 30 seconds of the onset of the quake, the waterfront had begun to shake violently. At Berth No. 1 on the Seward docks, crane operator Dean Smith noticed the first shock. Sitting in his operator's cabin some 50 feet above the pier, he felt the gantry crane begin to shake wildly, whipping back and forth with the motion of the earthquake. The wheels at the base of the crane came off the tracks and the other crane on the docks tumbled into the bay. He described the motion of the crane as "walking around like some stiff-legged spider." Smith quickly climbed down and ran to safety.

Meanwhile, the coastal tanker *Alaska Standard* was at Standard Oil Company's fuel dock taking on various petroleum products. The ship was tied to the dock by seven mooring lines, and five fuel hoses were hooked up and in use. Two grades of gasoline and stove oil had already been loaded, and diesel fuel was now being pumped aboard. Seaman Theodore Pedersen was on "hose watch" on the dock when the earthquake struck. What followed can only be described as incredible and qualifies Pedersen for nomination as one of the luckiest men in Alaska.

During the earthquake the ship "bucked" and then slammed against the pier, and Pedersen began to run up the dock toward shore. As he was running, pilings began shooting into the air around him and the 200-foot-long warehouse alongside the dock began to sink down as entire sections of the Seward waterfront started sliding into the bay. Fuel from 14 storage tanks, ruptured by the violent shaking, began to spread across the water. Then the *Alaska Standard* heeled sharply away from the pier, breaking the fuel hose connections and causing "geysers of oil to shoot skyward." Pedersen had run about 100 feet up the dock when the fuel caught fire and the Standard Oil Company tank farm blew up in "a big ball of fire." Next, according to the ship's master, Captain Solibakke, the tanker dropped vertically 20 to 30 feet, hitting bottom, and then "jumped straight up in the air," landing on the dock.

As the dock was collapsing, Ted Pedersen fell into the water and was struggling to stay afloat, some 20 feet below the main deck of his ship. He looked up just in time to see a huge wave, filled with debris from the dock, coming down on top of him. He was struck on the head and then lost consciousness.

This 30-foot-wave was a local tsunami generated by the slumping of the Seward waterfront into the bay. A small tsunami wave now carried flaming oil to the nearby Texaco petroleum facilities, setting them afire. It was truly a scene from hell. The *Alaska Standard* had been washed away from the burning dock by the wave, but the ship, with its cargo of gasoline and oil, found itself surrounded by flames on the surface of the water. With the ship now under power, Captain Solibakke managed to skirt the flaming water while heading his vessel toward the entrance to the bay. Meanwhile, the crew was busy fighting small fires, which had already broken out aboard ship. They discovered that one of their members, Donald Herrington, was missing and then saw a man floating astern, clinging for dear life to barrel. Alas, in their precarious situation there was absolutely nothing they could do, and the man soon disappeared from sight.

Seaman Pedersen, however, now regained consciousness. He was surrounded by debris and had a broken left leg, but he was alive—and he was lying on *Alaska Standard!* He had miraculously been washed aboard his own ship by the tsunami wave and had come to rest on a catwalk 8 feet above the ship's deck.

The wave performed another beneficial service. Along with Ted Pedersen and the mass of debris from the dock and warehouse that had been washed aboard the *Alaska Standard*, the Coast and Geodetic Survey tide gauge was also found. Though it was badly damaged, the record was later salvaged and showed that the water began to drop at 5:41 P.M., about 5 minutes after the onset of the earthquake—then the record becomes indecipherable.

By the time the local landslide tsunami wave had hit the waterfront area, most of the people in town had already left for higher ground. About 20 minutes later, the first wave of the major tectonic tsunami struck. This 40-foot wave must have been a terrifying sight, literally a wave of fire, as

Figure 6.3 Illustration depicting the fate of the *Alaska Standard* during landslide tsunami at Seward, Alaska, and the amazing survival of Seaman Ted Pedersen.

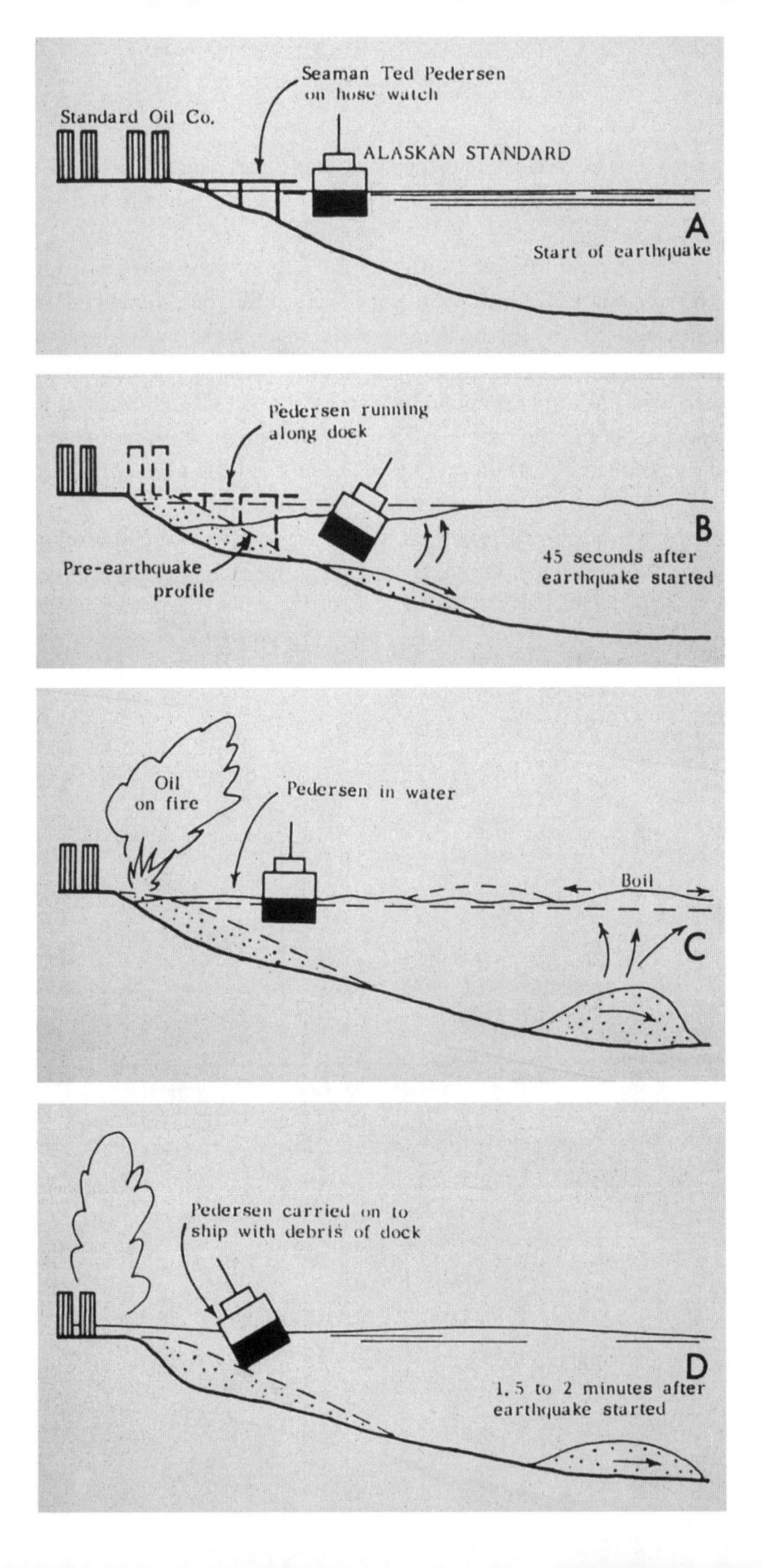
Seaman Ted Pedersen
on hose watch
Standard Oil Co.
ALASKAN STANDARD
A
Start of earthquake
Pedersen running
along dock
B
Pre-earthquake
profile
45 seconds after
earthquake started
Oil
on fire
Pedersen in water
Boil
C
Pedersen carried on to
ship with debris of dock
D
1.5 to 2 minutes after
earthquake started

it surged into Seward carrying the flaming oil. The Alaska Railway docks, the electrical generation plant, and many homes caught fire and were destroyed.

An 80-car train had been sitting in the railroad yard waiting to leave when the earthquake struck. The last 40 cars in the train were filled with gasoline and oil. When the flaming wave reached the tanker cars they began to explode one by one along the length of the train. The tsunami picked up a 120-ton locomotive and carried it over 100 feet, where it was left sticking out of the rubble. A 75-ton switch engine was carried nearly 300 feet and left upside down. One of the railroad engineers, Emil Elbe, died of a heart attack while running from the tsunami.

Not far from the railroad yard, Carl Christiansen and two brothers, John and Robert Eads, were looking toward the Standard Oil Company dock when they saw a 30-foot, fire-covered wave begin moving in their direction. Carl and Robert jumped into a car and John hopped into his pickup truck to try and flee the scene. The truck made it a little over 100

Figure 6.4 Damage to rail and port facilities at Seward, Alaska.

feet before the wave caught it, flipped it end over end, and carried it up a side road. John managed to climb out of the wreck and swim to safety through water filled with mud and debris. Meanwhile, the wave had also caught up with Carl and Robert in the car and washed them about 50 feet into the bushes off the side of the road. They too managed to climb out and were able to run to safety.

At about the same time, Gilbert Nelson and his family were driving toward Seward along the airport road. They stopped near the end of the main runway in time to see a gigantic wave moving directly toward them. The wave was not breaking but instead "rolling" and sounded "like a freight train." Boats, trees, and other debris were being borne up by the tsunami as it bore down on the Nelsons. They immediately turned their car around and began driving up an old runway at 70 miles per hour, with the wave in hot pursuit. Next the wave picked up the Civil Air Patrol building from near the end of the runway and began carrying it along behind the Nelsons. According to Mr. Nelson, there was "a great rush of wind just in front of the wave." Then their engine died. Fortunately they had gotten far enough ahead of the wave to have time to climb out of the car and run to safety.

Seward patrolman Dale Pickett had been driving toward the docks when he saw the Standard Oil tanks begin to explode. Before he could get out of his car, the tsunami wave had surrounded him and he was thrown against the side window, cutting his head. He tried in vain to open the doors, but the force of the water outside kept them shut. Finally the water penetrated inside the vehicle, equalizing the pressure, and he was able to open the door and flee to safety.

Soon after the tsunami struck, the Civil Defense team had assembled and gone to the waterfront. Fires were blazing, but there was little they could do—both power and water lines had been cut. In order to get water for the pumper truck, they had to drive a mile to First Lake and cut a hole in the ice. The Texaco tank farm would burn for two more days before the fire was finally brought under control.

A total of 12 people died at Seward as a result of the tsunami, and there were numerous close calls. A group of eight people had been forced to spend the night on the roof of a house floating near the airport. Fleeing the tsunami, they had first climbed up on the garage and then clambered higher onto the roof of the main part of the house. The garage had been carried away, and later waves reached the eaves of the main house, but they had managed to hang on to the roof. It snowed during the night, so they cut a hole in the roof and tore out insulation to keep warm. They had even tried four times to build a fire, but each time a wave splashed up on them, putting out the fire. Finally the house grounded itself in a grove of trees and they were saved.

At the warning center in Hawai'i, attempts were being made to obtain tide reports from stations near the epicenter in order to confirm the generation of a tsunami. But the combination of violent earthquake shaking and tsunami battering had left south-central Alaska without a single working tide gauge. In fact, the only good record of the tsunami along the Gulf of Alaska was made by personnel of the U.S. Navy Fleet Weather Station at Kodiak. By timing and marking the crest heights of successive waves and estimating the ebb levels and times, naval personnel were able to construct a record of the tsunami waves. This information was passed on to the warning center with great difficulty. The earthquake had knocked out all radio circuits in the building, but a telephone message was relayed to a remote navy radio station, which then sent the message to Honolulu via San Francisco. When the first wave flooded the building all electric power was lost, but within 15 minutes navy personnel had supplied emergency power and were able to send a follow-up message with more details about the first wave. Other than at Kodiak, the times and sequences of wave arrivals in south-central Alaska are known only from fragmentary eyewitness accounts. Most of the observers were naturally more concerned with saving their lives and property than with keeping track of changes in sea level.

Kodiak

Aside from the Seward and the Prince William Sound area, only the Kodiak Island group would experience heavy damage from tsunami waves. In fact, the Aleutian Islands and much of the Alaska Peninsula were largely shielded from the main tsunami by the Kodiak Islands themselves.

The city of Kodiak and the surrounding area on the northeastern side of the island form a major population center and even in 1964 counted some 4,200 residents. Following the earthquake, the mayor of Kodiak met with the police chief and city manager to discuss what action to take. They were aware of the potential danger of a tsunami but took no action until after a shortwave radio message describing tsunami waves striking an isolated native village on the opposite end of the island had been received at police headquarters. Now it became imperative to issue some kind of warning, but the earthquake had left the area without electricity or telephone service, and the Civil Defense siren was out of operation. With no plan for such a situation, the mayor finally decided to have the fire trucks turn on their sirens. Many residents responded by going into the streets to see why the sirens were being sounded. As people came out to see what the problem was, police herded them up nearby Pillar Mountain. A Navy Shore Patrol truck and a police patrol car went into low-lying areas of the city to spread the tsunami warning verbally. But not everyone heeded the warnings. Ironically, many of the casualties would be local fishermen who had listened but did not sufficiently understand

Figure 6.5 Some of the more than 100 fishing boats destroyed, damaged, or missing in Kodiak, Alaska, during the 1964 tsunami.

the tsunami danger. Upon hearing the warning, many of the fishermen went to the harbor to try to save their boats. The first wave was fortunately a gentle flood followed by a gradual ebb, but it alerted many residents to the danger and they fled to safety. However, the second wave came crashing in as a 30-foot wall of water, washing 100-ton fishing boats over the breakwater and as far as three blocks into town. Though several fishermen rode out the waves, six died trying to rescue their vessels.

Outside the city of Kodiak, the Curry and Vosgien families had spent Good Friday on a family outing. When the earthquake struck, they decided to return to the Naval Station. On the way to Kalsin Bay, a tsunami wave surged over the road, blocking their path. They waited for the water to recede and, believing that the tsunami was over, started out again but soon found their way blocked by debris in the road. They left their vehicles to seek help, but as Maurice Curry and twelve-year-old Richard Vosgien walked down the beach road toward a house, they were swept up and killed by the next tsunami wave. At nearby Kalsin Bay, ranchers lost some 80 head of cattle and Jake Blanc's house was moved half a mile inland, but amazingly left intact.

Also headed toward Kodiak, but from an outing at Chiniak, were Airman Gordon Wallace, his wife, Arlene, and his son, Jackie. When they found the road blocked, they got out of their car and were struck by a tsunami wave. Gordon was knocked unconscious. When he regained consciousness, he began frantically searching for his family along the shore. He rode a log through the icy water until he came upon a fence line, which he followed to a local house, where he collapsed. Gordon had survived, but his wife's body was found in the car, and that of his son nearby.

In or near the city of Kodiak, a total of 15 people were killed by the tsunami and damage would total nearly $11,000,000. The town newspaper (the *Kodiak Mirror*) building was picked up and washed off its foundation and laid to rest 30 feet away toward a hillside. Alf Madsen's cruiser, the *Explorer*, was deposited on the newspaper's former site, and a car and truck were washed into the Elks' Hall. Interestingly, the second wave, which swept away the Standard Oil Company and Alaska Packers facilities, uncovered an old Russian wharf. The stone structure, built in the late eighteenth century, appeared to be intact and undamaged. The wharf of the Alaska King Crab Company was, however, totally demolished.

Mentioning the loss of the crab company, the *Kodiak Mirror* stated, "but the crab season had practically ended anyway"—indicative of the ever-practical attitude of Alaskans.

Kodiak Island Eskimo Villages

As might be expected, several Pacific Eskimo villages were struck by tsunami waves. What happened at two villages is of particular interest with respect to how isolated native communities reacted to the disaster. Kaguyak, a village of 45 residents, and Old Harbor, with some 200 inhabitants, were both located on the south coast of Kodiak Island. No roads connected these villages with any other community, and their only means of physical communication was by sea or air. They did, however, maintain regular radio contact with other villages and the major town of Kodiak.

Both villages, built along narrow strips of land just above sea level, were extremely vulnerable to tsunami damage. Rather surprisingly, there is no native lore among Pacific Eskimos dealing with tsunamis. There is, on the other hand, what might be described as a "tsunami saint" of the Russian Orthodox Church. In August 1970, Father Herman (1757–1837) was canonized at Kodiak. A rather colorful figure, the monk was believed to have lived in a cave on Spruce Island. He wore a reindeer jumper, and he slept on a hard bench with bricks for a pillow and no blanket. He ate little, and was said to have worn 15-pound chain fetters around his ankles. Among his saintly acts, one in particular is of interest. According to legend, a tsunami was approaching the village of Elovio Ostrof when the inhabitants ran to the good father for help. He is said to have taken an "image of the Blessed Virgin" to the shore and placed it on the beach. Here he knelt down and, after a short prayer, he turned to the assembled people and said, "Do not be afraid, the water will not rise beyond the spot where His holy image stands." And apparently the future saint's prophecy was fulfilled.

Though there had not been a tsunami within memory at either village, Kaguyak and Old Harbor were somewhat accustomed to earthquakes. When the shaking from the Good Friday quake became violent, people grabbed their children, turned off their oil stoves, and ran outside their homes. According to Katherine, a resident of Kaguyak: "I heard the house start creaking. I told my husband, 'I think we are going to have earthquake.' Just as I said that, I told him 'you might as well put the stove out and let's run outside, 'cause the house might collapse on us.' And we did." After leaving her home, Phyllis, another resident of Kaguyak, commented about the violent tremors: "All the people was just sitting on the ground. They couldn't stand up on that ground."

The situation at Old Harbor during the earthquake was more complicated, for the town had been troubled for some time by a religious conflict. Most of the native settlements followed the Russian Orthodox faith, but a Protestant missionary had moved to Old Harbor several years before, built a chapel, and begun converting souls. There was now constant tension in the village between those still adhering to the Russian church and those converted to the "new faith." When the earthquake struck, most of the men in the village were congregated in the pool hall playing pinochle, but four young men had borrowed a boat, the *Kiska*, and gone to the old abandoned whaling station of Port Hoborn. They had gone ashore to salvage lumber from the old buildings to build a coffee shop for the village when the tremors began. As the shaking became violent, their boat began "bouncing on the bottom of the bay." Before their eyes, the three-story whaling building collapsed with a terrific noise and rocks began falling from surrounding cliffs. They ran back, quickly untied the *Kiska,* started the engine, and began to head out of the bay. Noticing that the water was receding from the shore, they increased the speed to full throttle, "so the beach wouldn't catch up with us."

Meanwhile, back at Old Harbor there had been a rapid exodus from the pool hall. At the height of the quake people were running "like they were drunk," barely able to stand, and frantically looking for their spouses and children. Cans fell from store shelves, the village church bell began to ring, and then the quake subsided.

A quick survey showed that there was little damage. Houses were generally "messed up" and two chimneys had fallen, but recrimination began almost immediately, with members of the Russian church accusing the Protestants of provoking the earthquake. According to one Protestant, they "thought it was because I had turned away from the Russian Church [and] had caused it." Protestants countered, "no, it isn't that that caused the quake. You should think of the [Orthodox] children that have been throwing stones at the [Protestant] chapel."

At both villages cleanup activity started almost immediately, and radios were turned on to hear the news from other parts of the island. But no one in either village anticipated a tsunami. On board the *Kiska* an unsuccessful attempt was made to contact Old Harbor by radio. The men on board noticed that the "tide" was receding even more rapidly, but it seemed, thankfully, to be helping increase their speed in the direction of the village. Meanwhile back in Old Harbor, villagers had begun to congregate at the beach to "admire the speed of the water" as it withdrew. Children would throw a piece of wood into the water, "to see how fast it was going out with the tide."

At Kaguyak, however, some villagers had begun to suspect that the worst was not over. A villager named Max, who had a new outboard motor on his skiff, had run down to check on his boat, and he noticed that the water had begun to rise. As the rising water gained speed, Max quickly reconsidered his priorities and began running for the hill behind the village. Another villager, Victor, saw the water rising and told his family, "big wave, big water coming from the ocean." He began running through the village, warning everyone he saw.

Meanwhile, Joe Melovedoff, the village radio operator, had run back to his house and began to send a series of radio messages. These warnings would be among the most significant messages transmitted during the entire Alaskan disaster. Joe first called his wife's uncle in Old Harbor and suggested that the villagers there head for higher ground. Next he radioed the Shearwater Cannery, which relayed his message to the station at Chiniak, near the city of Kodiak. Chiniak sent the message to the Kodiak Naval Station, but due to the confusion at the Naval Station, the warning was not broadcast over Armed Forces Radio. Fortunately, many with radios in Kodiak overhead the message from Chiniak and helped spread the warning. Though this transmission occurred only 20 minutes before the tsunami waves struck Kodiak, it provided the only real warning and no doubt helped save many lives. The warning message from Honolulu would not arrive for yet another 2½ hours!

As soon as Joe's message was received in Old Harbor, it was spread throughout the village by the age-old method of running and shouting. Villagers grabbed their coats, jackets, and children, and they began climbing the steep mountain behind the village. Mothers were carrying babies and older children were helping younger ones. Evan Naumoff, aged eighty-six, reported that he was helped by his ten-year-old grandson. "[He] was just going to grab my hand and going to drag me. [I said:] 'You got to try to save yourself. If God let me, I will go. If I make it, I'll make it. You got to save yourself.' I told him to go ahead and try to save himself, before the tidal wave was coming. Never mind me."

Several men had stayed near their homes until the water actually surged into the village. As they quickly fled, climbing the hill behind Old Harbor, they stopped to help stragglers, generally the most elderly. Larry Matfay, the radio operator, stayed in his house by his radio transmitter as the water came into the village. At the last minute he went outside to watch: "She (the wave) stopped right at our house." Fortunately, the first wave was rather gentle and only flooded the back of the village before the water began to recede. Larry radioed the status of Old Harbor to Joe in Kaguyak.

Meanwhile, the men on the *Kiska* were riding in toward the village with the high water. On board was the chairman of the town council, Evon, who stated:

> We were coming through the channel toward the village, and that was when we started getting pretty scared. . . . It was a ghostly sight because there were no lights . . . and we couldn't see nobody on the beaches. No dogs, nobody. It was real scary. As we got closer to the village, someone hollered from on top the ridge and said, "The people are on the mountain." [We] could see them on the mountains. One of us let out a sigh of relief, "The people are still alive" . . . but we didn't know yet if our families were hurt or not. We just hurried to the mooring as fast as we could. [We] tied up the boat and rushed ashore. [We] ran to our homes. Everybody split up to find our families. I went straight home and grabbed a sleeping bag and a blanket and then went up the hill as fast as I could to find my family. [They] were OK on top of the hill. I gave them the sleeping bag and went down again right away to try and get some more stuff.

As at Old Harbor, there was no appreciable damage to Kaguyak from the first wave, but the people on the hill knew that there might be another wave. Once the water began to recede from the first wave, every able-bodied man and woman in the village, including Phyllis, who was pregnant, went down to the village to retrieve essential items they might need for the night. They grabbed sleeping bags, diapers, and extra clothing. One child asked his father to bring his kitten, and he did.

Then they watched the water withdraw out into the bay farther than anyone had ever seen. Said one villager: "And boy! you should have seen that water go out of that bay from where we were! . . . Cleaned that whole bay out."

At Old Harbor, Evon was on his way down to find a coffee pot when he saw the water starting to come back in. He stopped and waited for the surge, but the second wave was not as big as the first. He turned to his father-in-law and said, "there are usually three." The water was withdrawing so quickly now that the *Kiska* broke from her mooring and began drifting away from the village. Villagers dragged a dory down to the water and two men, Larry and Mike, headed out after the *Kiska*, now far away and rapidly drifting toward open water.

Meanwhile at Kaguyak as soon as the second wave began to recede, Joe had returned to his radio, and other men hurried through the village gathering more supplies for the night. There was concern about more waves, so Sally, up on the hill with the women, children, and elderly men, was given a flashlight and told to watch for incoming waves and flash a signal if she saw one.

It was at this time that the villagers at Kaguyak saw a flare in the sky above the mountainside opposite their village. The flare had been fired from a site about two miles away, where Los Angeles geologist Donald Wyatt and his wife were camped. Twenty-six-year-old Walter Cohen was sent to see what was wrong. As he ran through the village on his way to the campsite, he quickly stopped at the church to say a prayer. No sooner had Walter left the village than the third wave washed into Kaguyak, but it only reached the embankment at the edge of the village.

But at Old Harbor the third wave came in with great force. Men down in the village gathering supplies heard "clam shells rolling" and ran for the hill. Meanwhile, Mike and Larry had reached the *Kiska* and were now anchored in front of the village. As the water began to surge into the village, they quickly cut loose the lumber they had gathered at Port Hoborn and started the engine. In Larry's words:

> At that time we were watching the houses start collapsing, just all kinds of noise coming down. And the houses come right on us, right on the boat . . . and we pushing them off with the poles so they wouldn't roll us over in the deep water. Tide was running swiftly—25 or 30 miles [per hour] swift coming through there. Just rivers, lots of water come out. All the buildings come right on us. . . hitting us hard.

The villagers on the hill above Old Harbor estimated the wave to be about 30 feet high as it hit the village. According to Evon, "This tide took most of the houses off their foundations and swung them over the back of the village and then when the water started going out, it took most of the houses out to sea with it."

Meanwhile, Larry, who had managed to keep the boat upright, fending off houses as they drifted out, radioed Kaguyak and let them know that a "big one" had hit Old Harbor. Next Larry radioed the people on the hill and asked them to blink a flashlight three times if they were safe. Someone heard him on a transistor radio and gave the signal. For a few minutes it was relatively calm so Larry and Mike were about to start a fire to make some coffee, when the *Kiska* suddenly rolled on her side. The water, withdrawing for yet a third time, had left the boat aground. Larry correctly interpreted that as a sign "that water is going to come again." He and Mike jumped off the boat and began running up the beach.

Back in Kaguyak, Joe had returned to the radio in his home and had made contact with a nearby crab boat. They discussed the possibility of rescue but determined that the crab boat's skiff was too small to transport the villagers, so four young men were sent down into the village to retrieve a large dory that had been washed behind the church.

In the meantime, Walter Cohen had reached the geologist's camp and had unsuccessfully tried to convince the couple to stay on the hill, where they would be safe, rather than trying to cross through the low-lying village and join the people on the opposite mountainside. The couple would not listen and insisted on crossing to join the villagers. Wyatt even refused to leave behind his equipment. So, loaded down with his knife, hatchet, pistol and gun belt, flare gun, and heavy backpack, he began the trek down the hill with his wife and Walter. From the beginning it was difficult for Wyatt to keep up, and after a short while the couple began to tire. Walter was forced to stop and let them rest before they had even gotten as far as the village. It was now dark and almost impossible to see what was happening with the water out in the bay. According to Walter, "Before we go to that flat [where the village is located] I let them lay down. I was . . . watching the moon and I start hearing that funny noise. Way out I begin to hear it. And I began to get kinda shaky, and scared."

Joe, who had been back in his house on the radio, had fortunately started back up the hill when the wave began to surge in, but the four men who had gone after the dory were now on flat land by the church. Up on the hill, Sally could just begin to make out the wave advancing across the bay and she tried desperately to signal them with her flashlight. As the wave approached, Max, one of the men by the church, leaped into his own small skiff, while the other three jumped into the dory. One of the men in the dory described the wave as being "fifty or seventy feet high." Walter and the Wyatt couple were now attempting to cross a creek near the church heading toward the men in the dory, when the wave came up behind them. A second "wall of water" now came toward them from the opposite end of the village. Just as Walter and the couple were picked up by the wave, the men in the dory grabbed them and pulled them into the boat.

The two giant waves met turbulently at the front of the village, forming several whirlpools. Max in the small skiff was caught between two whirlpools, and he described his boat as being "shot like an arrow" across the water to safety on the shore behind the village. But the big dory, now with six passengers, was caught in the whirlpools and swirled helplessly in circles. As the dory began to drift near shore, the geologist suddenly grabbed his wife and threw her out of the boat and onto the shore, where she landed face down in the mud and didn't move. Walter immediately leaped from the boat onto a large piece of ice drifting toward the incoherent woman. He ran across the ice, picked up the woman, jumped ashore, and began running up the hill carrying her. This was much to his own amazement as Walter was a small, slightly built man.

Meanwhile, according to one of the villagers still in the dory, "She [the dory] was jumping up and down. The dory was tip over. The other guy [the geologist] I can't see him. I stayed in the water myself about five or ten minutes or so. Boy, it was cold! Ice. Grass. Woods. The water. I was in there." The boat had overturned, throwing the four men into the water. The geologist, wearing his cherished and very heavy equipment, had sunk beneath the swirling water and was never seen again. Of the heroic villagers who had come to his aid, two were also lost to the tsunami, but one would be saved. As he himself described it,

> Maxie was jump off. . . . He was pull me up. I was stuck in the ice and I can't move. I was get weak, you know. Just I thought, "I'm not going to make it alive. . . ." I start to get numbed. . . . My leg can't move. Hard to

> breathe too. . . . After a while I come up again. And I was thinking "Where I am?" I don't remember nothing after I come up from the water. Just like a drunk. I passed out. Maxie was holding me. And we walk off.

Max, safely on shore, had seen his friend in distress and rushed to pull him from the water just in time. Max helped him up the hill, while Walter continued to carry the geologist's wife to safety, though she protested that she wanted to search for her husband.

Meanwhile, the villagers on the hill had watched in shock as Kaguyak was destroyed. The fourth wave had lifted the church and the houses and carried them out to sea. The villagers were especially upset by seeing their church washed away and agreed that it had been the first building to go. One villager said: "And [the] fourth wave was about fifty foot. . . . And it washed over all the buildings. The church, when I see the church I was crying all over the place. . . . And the wave took it away from us. . . . Nothing left in that village. Everything all gone. . . ." The lay reader's daughter Sally stated, "We pray to God to forgive us. . . . We sang. We cry inside and were still singing. . . . The water was high. The waves were coming. And when we start to kneel down and pray, that is when the water went out. Skooo! Just big noise. . . . And when it went back out it took our houses along too."

The geologist's wife was given coffee and dressed in warm clothes. A heated discussion about going back into the village to look for her husband and the two missing villagers followed, but with only three able-bodied men left, it was finally decided that it was too dangerous to risk losing the remaining men.

At Old Harbor as the wave surged in, Larry and Mike were running toward the beach from their grounded boat. They said they "could feel it coming, the roar," and looking back they saw the *Kiska* floating again. The water caught up with them just as they were about to reach the school, and they had to struggle through waist-deep water to get to the building. Now the tsunami did its worst and, according to Larry, "finished the job." The houses whose doors were closed drifted off their foundations and out into the bay, while those with doors open simply filled with water and sank in place. Some houses might have resisted the water, but they were knocked off their foundations by floating houses carried by the withdrawing current.

Most of the villagers covered their eyes in grief, but the oldest women in the village watched and would later state that the houses looked "just like boats going out, just floating in the sea. We hear the crunch of the houses cracking up. It was a creepy feeling." The people at Old Harbor reportedly began "crying when they said the houses were going." The *Kiska* drifted among the floating houses, carrying away one of only two radio transmitters in the village. The other transmitter had been in Larry's house, which was now also drifting out in the bay. Before leaving his house, Larry had lighted a kerosene lamp on the table in his house. Amazingly, the house had remained upright and the lamp could still be seen sitting on the table, shining through the window. The villagers on the hill now watched the eerie spectacle: "Everybody seen that. They said [to Larry], 'look at your house, still that light on your house.' Finally she went down, without tipping over. That was really something to see there."

During the remainder of the night, the villagers at Old Harbor waited on the hill, fearing more tsunami waves. They listened on their transistor radios and heard rumors that "Anchorage was a sea of flame with 500 dead," and that "50 were dead in Kodiak"—both exaggerations. At one point they even heard a radio report that "Old Harbor was wiped out!" Though most the buildings in the village had, indeed been destroyed, there had been no loss of life. But with no radio transmitter to contact the outside world, they were unable to rectify the incorrect news of their demise.

During the night both villages were contacted by radio from crab boats offering to evacuate any survivors. At Kaguyak villagers replied by radio that it was too dangerous and asked the skippers to take the boats out to deep water and return in the morning. With no radio transmitter at Old Harbor, several men had to go down to the beach and literally shout a message out to a waiting crab boat.

The next morning, village men went down to survey the damage. The destruction was total. At Old Harbor one man commented that there was "Just a big hole where the houses used to be. [It was] all muck, smelled like a sewer." The outhouses and cesspools had been swept away, too, and sewage was exposed throughout the village. Others nearly in shock at the scene seemed almost unconcerned. One man's house was still standing, and he told the people to help themselves to anything they needed.

In Old Harbor a total of 30 homes were completely destroyed; 8 others were damaged beyond repair. But the Russian Orthodox Church had not been washed away. Watermarks were found high on the church walls, but it had withstood all four tsunami waves. The Orthodox members of the community spoke with reverence and awe of how "their" church had survived the tsunami, "the water came [but] . . . that is the only one that stays." They quickly added that the Protestant missionary's chapel had been washed away, but failed to mention the fact that the missionary's home and the schoolhouse had also withstood all four waves. In fact, even during the tsunami, Protestants had been accused of causing the disaster: "You should not have gone to the missionaries! You are making the water come up!"

Throughout the morning the people in the two villages waited for rescue. Those in Old Harbor clustered in two groups representing the competing religions, whereas at Kaguyak the villagers all stayed together, but the father of one of the missing men and their wives had to be restrained from leaving the safety of the hill—there was still some fear of further tsunami waves.

Just before noon, some 18 hours after the earthquake had struck, a seaplane with a U.S. Navy doctor landed at Old Harbor. Upon observing the villagers the pilot reported, "They were actually in a state of shock . . . they were still standing up on the side of the hill, most all of them."

The crab boats were now ready to evacuate the villagers. At Old Harbor, a dory with a working outboard motor had been salvaged by the men and was used to ferry people out to the boats. On the one of the trips out to the boats, a villager mentioned the debris in the bay: "It seemed funny, you know, to see everything floating around. A roof here and there. Nobody said anything. Just looking." By 4 P.M. everyone from Old Harbor was safely on board the crab boats, where conditions were now quite crowded and one of the villagers remarked, "I know how it feels to be a sardine."

The survivors at Kaguyak were also evacuated to a crab boat for the trip to Kodiak. During one of the trips out to the boat, they spied the body of one of the missing men. It was draped over the rafters of a gable roof in a way that indicated he had survived the tsunami wave itself but had later died of exposure during the night. To their tremendous loss was now added a feeling of guilt at the thought that they might have been able to save him.

Though Kaguyak and Old Harbor were both devastated by the tsunami, proportionately the greatest loss of life in any village in Alaska took place at the native village of Chenega. Located on the south end of Chenega Island in Prince William Sound the settlement was struck only 10 minutes after the earthquake by a tsunami wave surging up over 70 feet above sea level. Only the school and one house survived the impact, the remaining buildings including the church and the village store being completely washed away. Many villagers were caught in their homes or along the beach, including most of the church elders who had taken refuge in the church sanctuary. Of the 76 inhabitants of Chenega, the death toll lay at 23, nearly a third of the village killed by the tsunami.

At Port Nellie Juan, the water rose to nearly 70 feet above sea level and three lives were lost. In Orca Inlet near Cordova, the U.S. Coast Guard vessel *Sedge* ran aground when the water level dropped 27 feet between waves. At nearby Point Whitshed, ten cabins floated out to sea in a line like baby ducks. The owner of one of the cabins had returned, thinking the tsunami over, and he too was lost.

At the southwestern tip of Kayak Island the Cape St. Elias lighthouse was rocked by the earthquake but sustained little damage to the structure itself. However before the earthquake, Coast Guardsman Frank Reed had crossed the gravel bar to nearby Pinnacle Rock to photograph sea lions. About 5 seconds into the initial shock, a 200-by-400-foot rock slide roared down Pinnacle Rock into the sea. When Reed had not returned to the lighthouse by 6 P.M., his three fellow Coast Guardsmen went to search for him. They found Reed, his leg broken by the rock slide and unable to walk. Placing him on a stretcher, they were headed back across the low gravel bar toward the lighthouse, when at 6:16 P.M., exactly 40 minutes after the earthquake, the first tectonic tsunami reached Cape St. Elias. A 4-foot wave surged over the bar, leaving the men struggling in chest-deep water. Then only 10 seconds later, a 10-foot wave surged over the group, sweeping all four men into deep water. Frank Reed was drowned, but the other Coast Guardsmen managed to swim back to shore.

Meanwhile, the warning center at Honolulu had received the first message from Kodiak and they had begun to prepare a tsunami warning message for dissemination across the Pacific. A more complete tide report from Kodiak came in at 8:11 P.M. and the following Pacific-wide tsunami warning was issued at 8:37 P.M.: "A sea wave has been generated which is spreading across the Pacific Ocean. . . . The intensity cannot, repeat cannot, be predicted."

At 9:11 P.M., with communication finally restored, Sitka, Alaska, was able to report that the tsunami had first reached there more than 2 hours earlier at 7:10 P.M.

Canada

As time passed and the tsunami waves spread out from Alaska toward Canada, there was more opportunity for a warning. Ironically, the Canadian authorities had withdrawn from the warning system just the previous summer and there was no official warning provided to Canada.

The Good Friday earthquake had been strongly felt in western Canada, and as far away as Vancouver and Calgary hanging light fixtures were set swinging. In British Columbia, seiches, set in motion by the earthquake, caused chunks of ice to be pitched up on the surfaces of frozen lakes. At François Lake, holes in the ice turned into geysers as water squirted up into the air. Although the earthquake caused no damage in Canada, there would be devastation—all caused by the tsunami. Tsunami waves swept down the coast of British Columbia, where in many localities they would strike near high tide.

At 11 P.M. Pacific Standard Time the first wave surged into Nootka Sound—Esperanza Inlet, washing up the main street of the town of Zeballos and causing extensive flooding damage. At Winter Harbour in Quatsino Sound a 38-foot boat, which had been carried out of one inlet by the withdrawal of the first wave, was washed up another inlet by the second wave and left stranded on the beach. The tsunami even managed to surge far enough up the Fraser River to produce a distinct record on the water-level gauge at Pitt Lake, a tidally influenced freshwater lake over 30 miles from the sea.

But it was the southwestern coast of Vancouver Island that bore the brunt of the tsunami. Of the 20 homes in the Indian village at the head of Hot Spring Cove, 18 were washed off their foundations and carried out into the inlet. Ironically, the largest wave heights would be measured not along the open Pacific coast of British Columbia, but instead up Barkley Sound at the head of narrow Alberni Inlet nearly 40 miles inland. Here lay the twin cities of Alberni and Port Alberni, an industrial center known for plywood, pulp, and paper products. The dimensions of Alberni Inlet are such that its normal period of oscillation (seiche) is very close to the dominant period of tsunami waves reaching the continental shelf off Vancouver Island. Resonance between the period of seiche and that of the tsunami produce wave heights at the head of the inlet that are 2 to 3 times greater than those along the open coastline outside the inlet.

But fortune would smile on Alberni and Port Alberni. As the tsunami reached the mouth of the inlet, an alert lighthouse keeper telephoned ahead to the twin cities, giving them a 10-minute warning while the waves traveled the 40 miles up the inlet. In another bit of good luck, the first wave was rather small and gentle, serving as a warning to many residents. The Royal Canadian Mounted Police then went from house to house warning about the potential danger of the tsunami. An hour and a half later a second, much larger, wave struck. Merna Semple, a resident of Alberni, was having coffee with friends when a neighbor called to warn of the wave. "We could see it coming across an intersection about half a block away," she explained in a provincial government film titled *Tsunami.* "So we went upstairs and got our four children and came back down. By this time the second wave was coming. We opened the back door and the front door and just let the wave go through the house. That was the only thing that saved the house from getting moved around on the foundation." Other houses were not so lucky. As another Alberni resident reported, "I was standing in a foot of water after the first wave hit. Suddenly the second one surged up into the street. I heard people screaming and men running back and forth across the street in front of the wave. I was amazed to see two big houses, 30 feet by 50 feet and two stories high, floating out in the Somass River. They gradually broke up and sank." Many houses along the northeast bank of the river were washed more than half a mile up river at speeds estimated by observers at more than 20 miles per hour. Logs and lumber piled up for shipment were turned into waterborne battering rams, and fishing boats, torn from their moorings, further added to the mass of floating debris. The water flooding into town seeped into underground storage tanks at service stations, sending gasoline flowing into the streets. Fires quickly broke out as electrical short circuits ignited the gasoline. The highest wave reached nearly 21 feet above sea level. Though most residents were asleep when the light keeper sent his warning, there was miraculously no loss of life or even serious injury. The area Civil Defense commander later reported: “I am unable to account for the lack of casualties.” The damage estimate, however, would exceed $5,000,000.

Washington and Oregon

Continuing on their path of destruction, the waves next assaulted the coast of Washington state. At Ocean City a park ranger reported: “It came over the dunes shooting 5 to 6 feet high, tossing logs around like match sticks.” At the Copalis River, Leonard Hulbert had stopped his car on the bridge to watch driftwood piling up against one of the supports. The bridge suddenly collapsed, plunging him into the river inside his automobile. Hulbert suffered a broken arm but somehow managed to open his car door and swim to safety. The swirling water tore two spans from the bridge.

Near Gray’s Harbor, the David Smiths were sleeping in their frame house with their two grandchildren when the second tsunami wave ripped their house from its foundation and washed it into Joe Creek. The house was slammed against a bridge and two spans torn loose, but the Smiths would miraculously survive, suffering only minor bruises. At Long Beach, four boys camping were chased from their tent by the rising water—their car was washed away. A woman in Olympia was suddenly awakened by water rocking her trailer. She stepped out the door into waist-deep rushing water. At Aberdeen the tsunami was described by eyewitnesses as coming “with a terrible rush” without “any notification” and sending beach “logs flying around like toothpicks.” In contrast, at La Push the waves were described as a gentle rise in the water level, though boats and a floating dock were broken loose from their moorings.

In the seaside town of Cannon Beach, Oregon, the tsunami picked up drift logs along the shore and carried them into the school yard, damaging steel playground equipment. Had the tsunami struck during the day when the school was in session rather than at night, a horrible death toll could have resulted. A family of six camping at Beverly State Park on Depoe Bay near Newport, Oregon, would not be so fortunate. The tsunami caught them asleep in their sleeping bags on the shore—their four children were washed out to sea and drowned. Much of the damage in Oregon occurred away from the oceanfront along estuary channels, where homes, businesses, and bridges were destroyed or severely damaged. The size, shape, and depth of these estuaries apparently were the main factors in determining whether the tsunami was dissipated near their mouths or propagated upstream as dangerous waves.

California

The response to the tsunami in the state of California was, perhaps, the most unusual of any location in the United States. At the state capital in Sacramento, the California Disaster Office (CDO) had received the first tsunami warning message at 9:36 P.M. local time, but it had not disseminated the information to coastal areas. In fact, it was not until after the second and third messages were received that an alert was sent out, and not by the CDO, but by the California State Department of Justice. They sent out an "All Points Bulletin" directed to sheriffs, chiefs of police, and Civil Defense directors in coastal areas.

Just south of the Oregon border lay Crescent City, a coastal town named for its crescent-shaped bay and with a history of susceptibility to tsunamis. The county sheriff received the alert at 11:08 P.M. local time and immediately contacted Civil Defense authorities. They met at the sheriff's office at 11:20 P.M., but since coastal areas to the north, with earlier scheduled tsunami arrival times, had not reported destructive tsunami waves, they delayed taking action. They did not understand that Crescent City might respond to tsunamis very differently from the other areas. At 11:50 P.M., 10 minutes after the first tsunami wave, rising 14 feet above sea level, had struck, the sheriff's deputies and local police finally began going door-to-door in the waterfront areas to warn residents. But many local residents had experienced previous tsunamis, such as that of 1960, and assumed that the city would receive only a few mild surges. The operator of the Texaco station had heard the news that a tsunami was due to hit Crescent City. He told a tourist who asked about the danger that they had had false alarms before, and added that he had "laughed at several customers that asked me if I intended to close the station and get to safety." Their "assumption" seemed to be confirmed when a second wave, only 6 feet high, came in at 12:20 A.M. By now some residents, as well as curious tourists, had returned to coastal areas, to "watch the tsunami." Police tried to keep the tourists out in order to prevent possible looting, but they continued to allow residents and business owners into the inundation zone in order to begin cleaning up.

At 1:00 A.M. a third wave crested at 16 feet; it was followed by an exceptionally large withdrawal. This should have served as a warning that much worse was to follow. The interval between the first three waves had been about 40 minutes. If this pattern held up, the next wave would arrive at about 1:40 A.M., just after high tide. No one in Crescent City knew that two mountains on the Pacific Ocean floor were already conspiring to destroy Crescent City, by causing the tsunami waves from Alaska to bend toward the town. Cobb seamount, 400 miles to the northwest, and another seamount over 100 miles west-northwest of the city were slowing the tsunami waves in such a way that the energy was being focused on Crescent City. The fourth wave began right on schedule at 1:40 A.M., continuing until 2 A.M., when it peaked as a mighty 21-foot surge. This wave moved through the coastal area of Crescent City as two deadly wedges. The wedges met and surged to the east across Highway 101 toward the Long Branch Tavern. The tavern's owner had been at home celebrating his birthday, but upon hearing news of the tsunami he had rushed down to the building to remove the money from the cash register. In fact, the party itself had moved to the tavern, and now eight people, thinking themselves safe from further waves, began to have a round of beers. A few minutes before the wave struck, a Coast Guard vehicle had driven by and a warning was shouted, but everything appeared normal and the party continued.

When the wave surged into the tavern, the party climbed onto pieces of furniture to stay above the water. Water continued to pour in until there was barely room under the ceiling to breathe. But then the water calmed as the wave crested, and two men swam toward shore to get a boat, while the rest of the group climbed up on to the roof. The men quickly returned with the boat, picked up the party on the roof, and began to row toward dry land, only about 75 feet away. They were almost to shore when the tsunami began to withdraw. The boat was sucked into Elk Creek and pulled toward the bay. As the boat was smashed against the steel grating of the highway bridge, one of the men managed to grab hold and pull himself to safety, but the rest were thrown into the swirling water. Only one was a strong enough swimmer to survive; the other five were lost to the tsunami.

These were not the only tragic deaths in Crescent City that night. One man was drowned when his trailer was picked up and flipped over by the tsunami. A woman with her three children was attempting to flee the waves when she was overtaken by the surge, and her ten-month-old son and three-year-old daughter were torn from her arms and drowned. Two couples, who had been having coffee before the large wave struck, were trying to flee in their car when the water caused the engine to stall. They attempted to escape on foot but were separated. One of the women drowned and the other was seriously injured, suffering a broken hand, two broken legs, seven broken ribs, and blows to her face and the back of her head. This woman's injuries illustrate the violent motion inherent in tsunami waves and the fact that they are usually filled with deadly debris. The fourth wave picked up cars, trucks, and logs, and smashed them

Figure 6.6 Photo of damage to automobiles and a house trailer taken at Crescent City, California, on the morning following the March 28, 1964, tsunami.

against anything in their path. A log crashed into the Crescent City Post Office, spilling the U.S. mail into the mud.

There would be many close calls that night. A seventy-five-year-old woman was rescued after having been pinned in her bed under the collapsed roof of her house. Her house had floated three blocks from its foundation. A couple floated down the street in their "unsinkable" Volkswagen. Back at the Texaco station, the owner and his wife had come to inspect the storage tanks for leaks, when the third wave surged in. The owner's car was put on the service rack to try to keep it out of the sea water. When the fourth wave surged in, the owner and his wife jumped into the car, while the operator raised them to the top of the rack. The fourth wave also picked up a gasoline truck parked at the Texaco station and slammed it through the garage door of the Nickols' Pontiac Building. An electric junction box just inside the door was knocked loose by the impact and a fire started. The fire destroyed the building and spread to the Texaco tank farm, where five gasoline storage tanks exploded and would continue to burn for three days. Amazingly the station owner, his wife, and the operator managed to flee to safety.

Survivors were stranded among enormous piles of debris left in the wake of the tsunami. Road graders and other heavy equipment were used to rescue people and carry them to safety. But Crescent City had suffered 10 deaths in the tsunami and a dozen injuries serious enough to require hospitalization. In all, 29 blocks of the city had been destroyed or badly damaged by the waves, including 172 businesses, 91 homes, and 12 house trailers, plus 21 boats sunk—totaling an estimated $16 million. The tsunami damage and death toll at Crescent City alone exceeded the combined affects of all previous tsunamis on record for the entire mainland U.S. coast.

The tsunami waves continued down the coast of California. At the mouth of the Klamath River, just south of Crescent City, two men had been eel fishing when they were picked up by the water and carried about half a mile upriver. They had managed to climb up onto a log, but when the water began to withdraw, they tried to swim for shore. Only one made it to safety.

In the San Francisco Bay area the tsunami caused extensive damage to yachts in Sausalito and San Rafael. In southern California, Marina del Rey was the hardest hit. Here, 450 feet of dock was washed half a mile up the channel. At Morro Bay one observer described the scene as the withdrawing water tore yachts from their moorings: "It was like someone pulled a plug from the bay."

Hawai'i

Meanwhile, tsunami waves were still heading toward the Hawaiian Islands. The islands had the distinct advantage of getting their warning directly from the Honolulu Observatory. As early as 6:25 P.M. Hawai'i Civil Defense had activated its emergency operating center. At 8:43 P.M., following the issuance of the tsunami warning, the decision was made to sound coastal sirens and to activate the Emergency Radio Broadcast System. At 9 P.M., the siren and broadcast systems were activated simultaneously in all counties. Fifteen minutes later, the warning center gave the following estimated times of arrival for the Hawaiian Islands: Kaua'i, 10:15 P.M.; O'ahu, 11:00 P.M.; Maui, 11:00 P.M.; Hawai'i, 11:15 P.M. The sirens sounded again at 10:00 and 10:30 P.M., with fixed sirens supplemented by police-car sirens.

The first wave report in the Hawaiian Islands came from the Coast Guard at Nāwiliwili, Kaua'i, which recorded a 1-foot wave arriving at 10:45 P.M. Later an 11-foot wave would roll into Kahului, and at Hilo the waves would exceed 12½ feet in height. But there would be no casualties in Hawai'i—the alert and the evacuation had been successful. Even property damage would be light, but then again there was little left to damage: the tsunamis of 1946 and 1960 had already destroyed most of the vulnerable areas. In Hilo, restaurants near the head of Reeds Bay were flooded and a sidewalk at the west end of the Waiākea Bridge over the Wailoa River was undermined and collapsed. In Maui there was damage to Kahului Harbor totaling over $50,000, but it was restricted to the waterfront area.

At 1:00 A.M. the warning center issued an all-clear bulletin stating: "the larger waves have apparently passed Hawai'i." The tsunami waves continued across the Pacific, but they would cause no further damage. In Japan waves only 3 to 10 inches high came ashore.

Meanwhile back in Alaska, the crab boats with the refugees from Kaguyak and Old Harbor were making slow progress toward Kodiak. They were forced to steer carefully through seas charged with debris and had stopped along the way to pick up other survivors. By Sunday morning the last of the refugees had finally reached Kodiak, where they were taken to the airport and put in the care of Red Cross social workers. For people used to the out-of-doors, this would be the beginning of a long, painful experience living in confinement. It would be months before the Old Harbor villagers would return and reestablish their village. The village site at Kaguyak was to be abandoned and the villagers resettled in Akhiok and Old Harbor.

Mother Nature was not yet through with Kodiak. Less than a week after the tsunami, a severe storm with winds registering more than 75 miles per hour struck the town. All 18 of the fishing vessels that had survived the tsunami, virtually the entire remaining fishing fleet of Kodiak, were sunk by the storm.

In all, the Prince William Sound earthquake of Good Friday 1964 had resulted in 131 fatalities and between $400 and $500 million in damage. It is thought that the shallow focus of the earthquake (only 14 miles deep) contributed to the heavy destruction. An interesting aspect of the earthquake was that the epicenter was located on land, making this the first known earthquake with a continental epicenter to be associated with a major destructive tsunami. The tsunami was generated because vertical tectonic displacement took place over a large area of the sea floor.

Of the casualties, Civil Defense estimated that 119 were victims of the tsunami and only 12 of the earthquake itself. The death toll could have been much higher. The earthquake occurred when schools were closed and business areas uncrowded.

Ironically the fact that the tsunami struck the coastal communities of south-central Alaska at low tide, rather than high tide, may have actually increased the loss of life and property. Of the tsunami death toll of 119, as many as 82 were victims of local tsunamis, with the remaining 37 lost to the major tectonic tsunami. Many of the local tsunamis were caused by submarine landslides of poorly compacted sediments. During low tide the bearing pressure on these sediments was increased, heightening the likelihood of failure. In other words, though the inundation by the tsunami waves would have been greater at high tide, some of the disastrous local tsunamis might not even have occurred had the tide been higher.

California was, perhaps, the only place where the death toll could have been much smaller. There was adequate warning and time for preparation, yet the list of casualties was greater than any other place but Alaska, with 10 killed and 35 injured in Crescent City. The problem lay partially in the response to the alert and in public understanding of tsunamis.

Reaction by county and city civil defense organizations varied considerably. In Humboldt County, the second advisory issued by the warning center was received at 11:08 P.M. at the county sheriff's office. All agencies were mobilized at 11:18 P.M., evacuation of all persons in danger areas was completed by 11:40 P.M., and road blocks were established.

At San Francisco immediately upon receipt of the advisory, attempts were made to evacuate coastal areas. While 2,500 people were evacuated, an estimated 10,000 people jammed the beaches in hope of "watching the tsunami waves" arrive. In San Diego, attempts to evacuate the beaches were rendered useless by curious onlookers. Los Angeles County *made no attempt whatsoever* to evacuate the waterfront. If large waves had struck these areas the casualty lists could have been truly horrible.

Why had the waves from this giant earthquake been so small in Hawai'i? Experts believe that the geometry of the fault caused maximum energy to be projected perpendicular rather than parallel to the fault. The northwest coast of North America lay at right angles to the fault and received the largest tsunami waves. Luckily for the Hawaiian Islands they were not in the direction of maximum energy propagation.

But Alaska had not been so fortunate and a glaring gap was revealed in the warning system: there had been no procedure for warning of locally generated tsunamis. As a result of the Good Friday earthquake, the Alaska Regional Tsunami Warning System (ARTWS) was established in 1967, in order to provide timely tsunami watch and warning to Alaska for locally generated tsunamis.

The ARTWS originally consisted of a main observatory at Palmer, 40 miles north of Anchorage, and two secondary observatories at Sitka and Adak. The system has since been renamed the West Coast/Alaska Tsunami Warning Center (WC/ATWC), and only the headquarters at Palmer is still in operation, but the WC/ATWC now has responsibility as the tsunami warning center for Alaska, British Columbia, Washington, Oregon, and California.

Because of the speed with which a locally generated tsunami can strike the various areas of the Alaskan and west coast region, the WC/ATWC must react very quickly to all earthquakes that could possibly generate tsunamis. As a result, all personnel are required to live within 5 minutes travel time to the center and have alarms in their residences. Staff carry special alphanumeric pagers on them at all times. If there is a large earthquake in the Pacific anywhere between Kamchatka, Russia, and the tip of southern California, the center swings into action. An automated system has already begun determining the epicenter and magnitude of the earthquake and displaying this information on the pagers while the geophysicists are rushing to the warning center to confirm the analysis of the data. If the earthquake has a magnitude greater than 7.0, a tsunami warning is issued for Pacific coastal areas of Alaska, Canada, and the United States within two and three hours tsunami travel time from the earthquake epicenter. Other geographical areas, outside the warning area, are placed in a "tsunami watch" status.

Next the tide stations nearest the epicenter are monitored for the existence of a tsunami. If a tsunami has been generated, "watch" areas are upgraded to "warning" status. The warning and other emergency information are sent out by the Emergency Alert System, VHF radio, commercial telephones lines, Coast Guard and Marine Weather HF radio, the National and Alaskan Warning Systems, National Weather Service Weather Radio, NOAA Weather Wire, and military communication channels. The WC/ATWC also works with the Pacific Tsunami Warning Center in Hawai'i to help monitor tsunamis generated outside the north Pacific, but in order to avoid confusion, serves as the sole source of tsunami information for its region of responsibility.

As experience in Hawai'i and California has shown, it is not enough to simply warn the population. The public must understand the warning and respond. With this goal in mind, the WC/ATWC maintains a community preparedness program designed to educate the public about what to do if caught in a violent earthquake or tsunami. The program goes to each community and presents a detailed briefing covering the seismicity of their area, past historical earthquake/tsunami damage, and estimates of what might happen if an earthquake or tsunami were to occur. Tsunami education programs are now being implemented or expanded in coastal regions of Washington, Oregon, and California.

Since its inception, the West Coast/Alaska Tsunami Warning System has responded to an average of more than a dozen alerts each month. It has done an excellent job of warning residents of the seismically active Alaska region about the dangers from local earthquakes and tsunamis.

Following the 1960 Chilean tsunami, the Pacific Tsunami Warning Center in Honolulu had been expanded as new members joined the system. An International Coordination Group was soon set up to review the activities of the warning system, and in 1965, the International Tsunami Information Center began working under the auspices of UNESCO.

With the headquarters of an improved and expanded warning system in Honolulu, would there ever be a need for a special local warning system for the Hawaiian Islands? Could Hawai‘i, too, be in danger from a locally generated tsunami?

7

Local Tsunamis in Hawai‘i

Most destructive tsunamis are associated with earthquakes and are caused by tectonic displacement of the sea floor. Fortunately, most of the Hawaiian Islands, with the exception of the island of Hawai‘i itself, are not very seismically active. The island of Hawai‘i does have a large number of earthquakes, but most of these are small and cause little or no damage. About once a century, however, a very large earthquake does occur on the island. Such a quake occurred in 1868.

The Great Earthquake of 1868

On Thursday, April 2, 1868, a major earthquake struck the island of Hawai‘i. The quake was felt as far away as Kaua‘i but caused extensive damage only on the Big Island, where it was reported that every European-style building in the Ka‘ū district was completely destroyed. An eyewitness to the destruction related his experience:

> At 4 P.M. on the 2nd a shock occurred, which was absolutely terrific. All over Kau and Hilo, the earth was rent in a thousand places, opening cracks and fissures from an inch to many feet in width, throwing over stone walls, prostrating trees, breaking down banks and precipices, demolishing nearly all stone churches and dwellings, and filling the people with consternation. This shock lasted about 3 minutes, and had it continued three minutes more, with such violence, few homes would have been left standing in Hilo or Kau. Fortunately there was but one stone building in Hilo, our prison, and that fell immediately.

Figure 7.1 Map of the island of Hawai'i showing selected locations mentioned in stories of the 1868 and 1975 earthquakes and tsunamis.

The earthquake, which probably had a Richter magnitude of between 7.25 and 7.75, triggered a large landslide in Wood Valley on the slopes of Mauna Loa, about 5 miles north of Pāhala. Here a large mass of lava rock slid down slope over a layer of wet ash, going from an altitude of 3,500 feet down to 1,620 feet. In all, the slide covered a distance of some 2½ miles and completely destroyed a Hawaiian village. Thirty-one natives, and more than 500 horses, cattle, and goats, were buried alive. As if the earthquake were not enough, a tsunami was generated. A Mr. Stackpole, returning from the Volcano House at Kīlauea to the shore at Keauhou, met the men who worked at the Keauhou landing running up hill. They reported that immediately after the earthquake the sea had rushed in and "swept off every dwelling and store house."

According to another account, the tsunami "rolled in over the tops of the coconut trees, probably sixty feet high, and drove the floating rubbish, timber, and soforth, inland a distance of a quarter of a mile in some places, taking out to sea when it returned, houses, men, women, and almost everything movable. The villages of Punalu'u, Ninole, Kawaa, and Honuapo were utterly annihilated." It is now estimated that the tsunami waves had a run-up height of about 45 feet at the Keauhou landing and 9 feet at Hilo. This deadly local tsunami resulted in the loss of 46 lives in Hawai'i. The waves spread out across the Pacific, registering on tide gauges in Oregon and California some 5 hours later. But outside of Hawai'i, the waves were very small—measuring only 4 inches at San Diego—and caused no damage. The 1868 tsunami produced one of the most fantastic of all stories about tsunamis. This is the tale, as related by Mr. C. C. Bennett:

> I have just been told of an incident that occurred at Ninole, during the inundation of that place. At the time of the shock on Thursday, a man named Holoua, and his wife, ran out of the house and started for the hills above, but remembering the money he had in the house, the man left his wife and returned to bring it away. Just as he had entered the house the sea broke on the shore, and enveloping the building; first washing it several yards inland, and then, as the wave receded, swept it off to sea, with him in it. Being a powerful man, and one of the most expert swimmers in that region, he succeeded in wrenching off a board or a rafter, and with this as a *papa hee nalu* (surfboard) he boldly struck out for the shore, and landed safely with the return wave. When we consider the prodigious height of the breaker on which he rode to the shore, (50, perhaps 60 feet), the feat seems almost incredible, were it not that he is now alive to attest it, as well as the people on the hill side who saw him.

Both the epicenter of the 1868 earthquake and the site of generation of the tsunami lay near Kalapana, off the southeast coast of the island of Hawai'i. Nearly a century would pass before this area would again produce a large earthquake. On November 29, 1975, it happened again.

The Halapē Tragedy

Thanksgiving weekend. Time for a break. What better way to spend it than to go camping at the remote beach park of Halapē on the island of

Figure 7.2 Holoua surfing the tsunami following the 1968 earthquake along the south coast of the island of Hawai'i.

Hawai'i? One of Hawai'i's most idyllic retreats, Halapē lay just seaward of the base of the 1,000-foot cliffs of Pu'ukapukapu (Forbidden Hill). Because it could be reached only by foot or on horseback, Halapē remained an unspoiled spot in paradise. Thirty-four people, including several fishermen, a group from the Sierra Club, four hikers, and Boy Scout Troop 77, decided to enjoy the special pleasures offered by Halapē on that holiday weekend in 1975. The campers from Troop 77 included six scouts accompanied by four adults. They hiked in on Thanksgiving Day and settled down to have a good time. The four men had been looking forward to the trip even more than the boys. Dr. James Mitchell had arranged his schedule so that he could go with Claude Moore and Don White, and policeman James Kawakami had flown in from Honolulu that morning, directly after finishing his night shift. When they arrived at the beach, the four men set up camp in one shelter while the boys selected a spot in the coconut grove. It rained that first night, so the boys—David White (Don's son), "Fal" Allen, Mike Sterns, Leif Thompson, Noel Loo, and Timothy Twigg-Smith—moved to a second shelter about a quarter of a mile along the beach. The rain also created somewhat of a problem for the adults, as it caused the water tank to overflow into the shelter where they

Figure 7.3 Aerial photo of the Halapē, Hawai'i, palm grove prior to the 1975 earthquake and tsunami.

Figure 7.4 Boy Scout Troop 77 on their hike down to Halapē for Thanksgiving Day weekend, 1975.

were sleeping. The water soaked their shoes, and so they hung them from the roof to dry.

The next day was spent fishing, swimming, and being lazy. That was the essence of a camping weekend. Keith Stratton and his friends Brad, Barbara, and Beverley, who had also hiked in, picked opihi and shared it with the fishermen and the Boy Scouts. They rinsed off in brackish water held in a small crevice in the rocks before going to sleep on the sand. They had decided not to sleep in the coconut grove in case coconuts fell on them! That night the scout group went to sleep, situated as they had been the night before, except that David White decided to pitch his tent in the coconut grove. At 3:36 A.M. the island was jolted by an earthquake. The quake, which measured 5.7 on the Richter scale, was centered beneath the south flank of Kīlauea Volcano, about 3 miles inland of Kamoamoa.

At Halapē the campers were startled awake by the movement of the earth. Hearing the crashing sound of rock falls, they looked toward Pu'u-kapukapu. In the weak early morning light, they saw dust clouds rising as rocks fell down the steep cliff face.

After chatting about the quake for a while, some of the campers went back to their shelters to sleep. Others, however, decided to avoid the danger of landslides and moved closer to the ocean. They failed to realize that the main danger was to come from the sea. David told his father that he felt something was going to happen and that he wanted to leave. He was reassured and went back to sleep .

Little more than an hour later, at 4:48 A.M., a second earthquake struck. This quake was also beneath Kīlauea's south flank, but it was centered southeast of the Waha'ula *heiau* (a traditional Hawaiian temple), about 2 miles offshore, and unlike the earlier quake, it did not subside in intensity. The campers on the beach at Halapē experienced a terrible shaking and shuddering. Many parts of the island were plunged into darkness as the violent trembling felled utility poles and caused power outages. The quake was so severe locally that the seismographs at the Hawaiian Volcano Observatory went off scale. A Richter magnitude of 7.2 was later calculated by scientists at distant stations in North America, Japan, and New Zealand, where the seismographs had remained on scale. The last earthquake of this magnitude in Hawai'i had been the great earthquake of 1868, and that had been accompanied by a tsunami.

At Halapē, only 15 miles west of the epicenter, Claude Moore and his companions awoke to find the shelter shaking and the water tank rocking and splashing water. Worried that the water tank might collapse on them, they ran outside. Although able to stand when the quake first started, they were now thrown to the ground by the violent shaking. Next they heard a terrifying roar as boulders crashed down the cliff face of Pu‘ukapukapu.

At this time Keith Stratton and his friends began scrambling toward the sea, away from the threat of falling rocks. Brad shone his flashlight toward the ocean, and they saw first a huge hole, then the cresting wave. The water hit them and separated them. Keith was propelled through the scrub about 50 feet inland.

Claude had run back into the shelter to get his shoes. While he was fumbling to detach the laces of his shoes, which had been tied to the roof to dry, he could hear the other adults in his party talking outside with two campers, who had just been drenched by a wave as they slept on the beach. Claude heard someone say something about water, and he thought of the possibility of a tsunami. All of a sudden there was silence. He ran out of the shelter toward the beach to see what was happening.

Claude looked to the right and saw a group of people running as fast as they could away from the beach toward Pu‘ukapukapu. He looked back to his left and saw a 5-foot wall of water crashing through the coconut palms. It was almost upon him. He flung himself behind the 4-foot stone wall at the back of the shelter and hung on for all he was worth. At first he was doing well, holding on to the top of the wall and floating on the surface of the water. Then the pressure of the wave proved too much for the stones, and the wall started to collapse. At the same time the shelter was swaying and creaking, so he let go. "This is it," he thought, as he was washed back and forth in the water. It seemed that he was swirled in the water for an eternity—although it may have been only a few minutes. Suddenly he was on his feet; the water was waist-high. A ray of hope penetrated the mist of his confusion. Perhaps there was a possibility of escape after all! A great blow from behind drove all thoughts from Claude's mind. The second wave, some 25 feet high, overwhelmed him. His face crashed onto a rock, stars flashed behind his eyelids, and the water was over and around him—then it was gone.

Claude found himself at the base of a crevice, some 20 feet deep and 30 feet across. He clambered up the back of the crack, realizing with a sense of wonder that he was still alive. Battered and abraded, his shirt hanging off his arms by the cuffs and his watch torn from his wrist by the wrenching wave, he was dazed and bewildered—but he was alive. He had no serious injuries, and began to wonder about the others. He stood up and tried to walk but became sick and dizzy. Once he heard Don White calling for David, and he called back to him. After that he heard no sound but the sea.

Further inland, Keith Stratton had been lucky enough to be left behind when the wave withdrew. He thought he was the only one of his foursome to have survived until he followed the sound of voices and found Brad, Barbara, and Beverley, shaken but alive. Together they waited for sunrise.

Just as the sea had destroyed the shelter to which Claude had been clinging, so it ravaged the structure where the Boy Scouts were sleeping. The arrival of the wave was the first they knew of their danger. The rushing water tore the shelter apart, pushing Timothy through the wall. All the boys became entangled in bushes and pulled to and fro by the water, until the waves receded and they were able to run to higher ground. There they waited until daybreak with three fishermen from Mountain View.

The group of 11 fishermen had ridden into Halapē on horseback and camped near the animal corral at the back of the beach. After they were wakened by the second earthquake, they heard sounds of a landslide coming from the cliffs and they ran toward the sea. They immediately reversed their direction, however, when they heard someone shouting, "Tidal wave!" They ran before the advancing wave together with several of the members of the Boy Scout party. They ran until they were confronted by a deep ditch, 8 to 10 feet wide. There was no choice, nowhere else to go. One of the fishermen, Michael Cruz, hesitated for a moment and disappeared forever into the sea. Some jumped into the ditch, and the wave followed, crashing on top of them. When they tried to climb out, more water would pour onto them, tumbling them over and around. According to one survivor, it was like being "inside a washing machine." The second wave, higher and more turbulent than the first, had washed everything in its path as far as 300 feet inland. Trees, debris from the shelters, rocks, and people were deposited in the ditch. Several smaller waves washed over the exhausted victims, but their uncomfortable refuge saved them from the fate of being carried out to sea. All those who had been in the ditch survived, although James Kawakami and Don White had swallowed a great quantity of seawater and sand. Both Don and David had been pounded by the water and debris, but David was not hurt.

After the waves had receded, David took care of his father. He pulled

Don into the shelter of a rock, covering his body with long grass and then lying over him to provide his own body warmth. When day dawned, Leif Thompson, who was not hurt, went searching for the others. He found Claude and reported to him that all members of the Boy Scout party were safe except for Dr. Mitchell, but that Jimmy Kawakami was in a state of shock after his ordeal in the ditch. Leif returned to Jimmy and stayed with him until help arrived.

Both Leif Thompson and David White received awards from the Boy Scouts of America for their life-saving efforts during the disaster. All the campers were rescued later that day. Keith Stratton's friend Cliff, who knew they had been camping at Halapē, flew in his plane over the area at daybreak and alerted the National Guard. Everyone was airlifted out by army helicopter, and the injured taken to Hilo Hospital. When it was their turn to be airlifted Keith's party were awed to see that the small crevice where they had bathed the day before had widened by almost 30 feet. Nineteen of the campers had been injured, seven requiring extended hospitalization. Four of the ten horses were lost. Michael Cruz had been taken by the sea and his body was never found. Of Claude's party, all were safe except for Dr. Mitchell. He had been battered and drowned among the rocks and debris of Halapē.

The locally generated tsunami that struck Halapē was caused by the sudden movement of the sea floor off the southeast coast of Hawai'i, the same movement that produced the 7.2 earthquake. Coincident with the earthquake, the ground along the shoreline subsided by up to 10 feet, submerging much of the Halapē palm grove. It is thought that the first wave was caused by water along the shoreline rushing in to fill the cavity caused by the sudden subsidence. The second wave, which was much larger, resulted as the deep-water tsunami wave arrived from offshore.

Amazingly, the loss of life was confined to Halapē, but property damage was widespread. The tsunami had radiated out in all directions from its source, traveling several hundred miles per hour. At the small bay of Punalu'u, 20 miles southwest of Halapē, residents in their homes as well as several families camping near the beach were awakened by the earthquakes. What no one suspected was that a tsunami was headed toward them. Many residents and campers were forced to wade to higher ground as the sea level rose steadily and quickly. But the big wave would not arrive for another 10 minutes. Cecil S. Carmichael and his wife were

Figure 7.5 Aerial photo of the Halapē palm grove after the 1975 earthquake and tsunami.

wakened by the foreshock at 3:38 A.M. but remained in their A-frame house in Punalu'u. After the big quake at 4:48, the Carmichaels got dressed and decided to go downstairs and drive away to higher ground. Their account of the tsunami is as follows:

> Before we could leave the house, we heard a sound like a strong wind. My wife pulled back the front drape and saw water over our deck and coming in under the front sliding door. This was at 5:00 A.M. Our lot is +9 feet and the house was 4 feet off the ground; i.e. the floor level was at +13 feet.
>
> We left by the back door using flashlights. The back steps were gone—also our 3000-gal. water tank and garage. We waded through knee-deep muddy water and climbed over wet stone walls and walked quickly in 1 foot-deep water to the main paved road, going through Dahlberg's yard. There the lei stands had been smashed back about 30 feet and a power line was down on the pavement. We heard another sound like a strong wind. My wife ran, and I shone my light and saw the second wave by the old wharf. This was at 5:10 A.M. I ran. When I reached the new road, I heard the crashing of metal and wood. The wave reached the lawn of Arnold Howard's house. I later measured the wave height at two coconut

trees and two hala trees in our lot using a tape, a staff, and a hand level. The height was 25 feet above sea level.

The third wave which hit at 5:20 A.M., must have had a height of at least 13 feet but less than 25 feet. Our house dragged a coconut tree with it 100 feet inland and broke apart—a total loss. We could see most of it afterward floating offshore with other debris; fishermen friends have told us they subsequently found lumber from it washed up at several locations between Punaluu and South Point. Our four stone walls were completely washed away. Our car by the wall was turned around 180° and was a total loss.

Throughout Punalu‘u, beach houses and oceanfront properties were swept from their foundations, and the Punalu‘u Village restaurant and gift shop were inundated. Although the restaurant structure remained intact, interior damage would amount to nearly $1 million; the waves left the floor covered in mud and the furniture jumbled and broken. Throughout the area, evidence of the advance of the sea could be seen in the form of stranded sea urchins, eels, fish, and starfish. In all, 7 homes and 2 vehicles were destroyed by the tsunami, but no injuries were reported.

Figure 7.6 House at Punalu‘u washed off its foundation by the tsunami of 1975.

Farther down its path of destruction, the tsunami claimed park facilities, a warehouse, and a fishing pier at Honu'apo. At Ka'alu'alu Bay, 15 miles southwest of Punalu'u, the waves damaged several vehicles and ravaged a campsite, terrifying seven campers. Although near the epicenter of the earthquake the waves were from 20 to nearly 50 feet high, they diminished in height rapidly as they spread away from their source.

The big earthquake had shaken all of Hilo. Some buildings had been structurally damaged, chimneys were toppled, windows were broken, and merchandise was knocked from store shelves. The edge of the parking lot at the Bayshore Towers condominium had fallen into the bay.

The first tsunami wave reached Hilo 20 minutes after the big quake. Although this wave was only 1½ feet high, it was followed by a major recession of the water some 5 feet below normal. At tiny Radio Bay, located in the corner of Hilo Harbor, the crew members of the Coast Guard cutter *Cape Small* watched helplessly as their ship settled into the mud and began to list to one side. Then the sea surged back in as the second and largest wave crested in Hilo Bay at about 5:30 A.M. This wave, up to 8 feet high, ripped small boats from their moorings, washing some onto the pier and sinking others. An automobile was swept off the pier into the harbor. In one peculiar incident, a man was thrown from his boat onto the pier by the advancing wave, then washed off the pier into the water, and back onto his boat as the wave receded. A series of progressively smaller waves continued to surge in and out of Hilo Bay at approximately 15-minute intervals for several hours.

The tsunami continued to spread out across the Pacific. The waves arriving in Honolulu 30 minutes later were about 1 foot high. The tsunami measured 1 foot in Los Angeles, where it was recorded 6 hours and 45 minutes after the earthquake, having crossed the Pacific from Hawai'i at a speed over 350 miles per hour. The waves took some 8 hours to reach Japan, where the largest wave measured almost 2 feet. In all, at least 39 tide stations outside of Hawai'i recorded the tsunami.

But death and destruction from the tsunami occurred only on the island of Hawai'i. The run-up height of the largest wave was 47 feet at Keauhou Landing, 26 feet at Halapē, 25 feet at Punalu'u, and 8 feet at the Wailuku River in Hilo Bay. On the island of Hawai'i, damage from the earthquake and tsunami exceeded $4 million. A total of 8 houses, 3 business, and 27 fishing boats were destroyed or severely damaged by the tsunami. Two people were killed, both at Halapē. The president of the United States declared the County of Hawai'i a disaster area.

How had the authorities reacted to this locally generated tsunami? Had there been time for a warning? The first reaction to the tsunami occurred when reports of unusual wave activity in Kona reached the Hawai'i County Police. They immediately ordered the evacuation of the Hilo waterfront. State Civil Defense ordered the coastal sirens to be sounded on Hawai'i and Maui. But the first wave arrived in Hilo almost 30 minutes before the sirens sounded. Coastal sirens on Maui were not sounded until 7:20 A.M., over 1½ hours after the earthquake. The alert was canceled at 8:30 A.M.

The implications are clear. Had the tsunami occurred at a time other than in the early morning hours, when few people were at the harbor or along the shore, or had the waves been larger, the death toll could have been much greater.

When scientists later studied the earthquake and tsunami in great detail, it was determined that the tsunami was probably too large to have been generated by the earthquake alone but may have been caused, at least in part, by a submarine landslide down the south flank of Kīlauea Volcano. Had landslides in Hawai'i caused tsunamis before?

Tsunamis Generated by Landslides in Hawai'i

In the 1920s, geologists studying the south slopes of the islands of Moloka'i and Lāna'i discovered limestone boulders high above sea level. Similar deposits were also found on the southwestern side of Maui, on easternmost O'ahu, and on the western side of the Kohala Mountains, on the island of Hawai'i. These curious deposits were composed of the skeletons of coral and other shallow marine reef-building organisms. How was it possible that skeletons of marine organisms should be found high above sea level? These deposits were interpreted as ancient shorelines formed during the ice ages when sea level stood higher than at present. More recently, geologists studying the subsidence rates of the Hawaiian Islands determined that the southeastern islands in the chain were sinking too fast for high sea levels formed during the ice ages to be above present sea level today. What other explanation might there be for these marine deposits at high elevations?

Figure 7.7 A ridge of basalt boulders at the upper surface of the Hulopo'e Gravel on the island of Lāna'i.

In the early 1980s, two geologists, James and George Moore, of the U.S. Geological Survey (USGS) made a detailed study of one such deposit, the Hulopo'e Gravel on the island of Lāna'i. This deposit contains coral sand at elevations as high as 1,070 feet and shows signs of wave erosion up to nearly 1,200 feet. The deposit becomes thicker and contains larger-size particles as you approach the sea. The geologists could interpret this only as being marine material that had been ripped off the bottom in the near shore zone and washed up to these elevations by a giant wave sweeping over the flank of the island. Other geologic evidence suggested that this event probably occurred around 105,000 years ago.

In historical times Hawaiians had witnessed tsunami waves washing to heights of more than 50 feet, but nothing even close to the incredible 1,200 feet called for by the geologists. What could possibly have generated waves of such prodigious size?

A different question, which had plagued geologists for some time, was the sheer size of the cliffs around the islands. Huge cliffs are prominent throughout the chain from the Nā Pali coast of Kaua'i, to the Nu'uanu Pali on O'ahu, to the cliffs fronting Waipi'o and Waimanu valleys and above Kealakekua Bay on the island of Hawai'i. Many were just too high to have been formed by normal erosion processes. How had these huge cliffs formed? The answer to both mysteries would not be long in coming.

In 1988, the USGS completed mapping the Exclusive Economic Zone around the Hawaiian Islands, a vast area of ocean and sea floor whose resources are considered the property of the United States under the Law of the Sea Convention of 1982. A newly developed sea floor mapping tool, named GLORIA,[1] had been employed for the first time in studying the Hawaiian Ridge. The acronym GLORIA is used to identify an advanced sonar mapping system that can produce images of as much as 14 miles of sea floor on either side of a survey vessel and resolve features as small as a few hundred meters across. Using GLORIA, scientists made an astounding discovery about the sea floor surrounding the islands. Instead of being covered with layers of sediment washed off the islands over time, the ocean bottom is almost completely covered with deposits of material that have slid, slumped, and roared down the island slopes as avalanches of volcanic debris. These deposits cover more than 36,000 square miles of sea floor, an area some five times the size of the islands themselves. One expert estimates that there are as many as 68 major landslides, with some individual deposits over 120 miles long and composed of nearly 1,100 cubic miles of material.

Detailed study of the deposits has revealed that they occur in two basic forms, as "slumps" and as "debris avalanches." Slumps are tens of miles wide, may be several miles thick, and move slowly, although there may be sudden "adjustments," which can produce earthquakes. Debris avalanches are longer, thinner, and move much faster.

At Kealekekua Bay, above the spot where Captain Cook died, is a sheer cliff marking the Kealakekua fault. Movement along this fault during a 1951 earthquake resulted in serious ground cracking in the Kona area of the Big Island. The fault and cliff continue underwater along the west flank of Mauna Loa Volcano. The submarine slope here is as steep as 25°, marking the area from which giant debris avalanches had slid northward and westward onto the deep ocean floor. Lying further seaward at depths from 10,000 to 15,000 feet and covering an area 35 miles wide by 50 miles

[1]The acronym stands for Geologic Long Range Inclined ASDIC (Anti-Submarine Detection Investigation Committee).

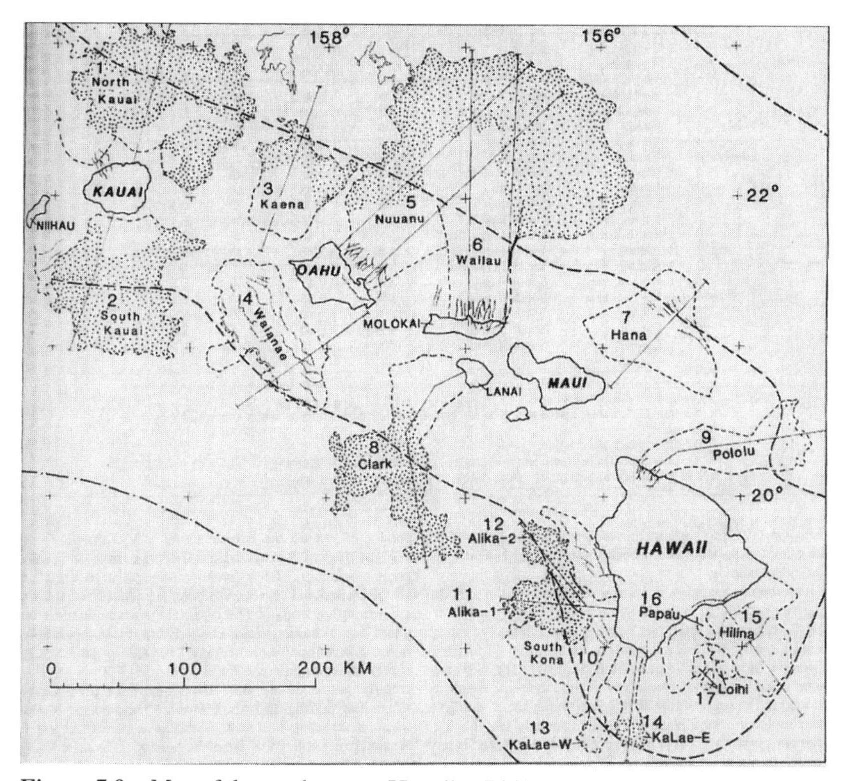

Figure 7.8 Map of the southeastern Hawaiian Ridge showing major slide deposits. Some 17 different slide deposits have been identified. Numbers 11 and 12 identify the 'Ālika slide, which may be responsible for the tsunami deposits found on Lāna'i.

long is a pile of volcanic debris as thick as 250 feet. This deposit, called the 'Ālika slide, is the material that slid off the west flank of Mauna Loa. Many geologists consider it as the most likely candidate to account for the giant waves that washed the deposits up on to Lāna'i and other islands around 105,000 years ago. Based on evidence of wave erosion occurring around this time in Australia, some researchers actually believe that the tsunami waves produced by this massive slide might have traveled over 4,000 miles to New South Wales, where they produced wave destruction to heights as great as 80 feet above sea level.

It has been estimated that the 'Ālika slide could have traveled along the sea floor at speeds as high as an amazing 40 miles per hour or more. Slumps may not travel with the high speeds of debris avalanches, but their results are just as earth-shaking. The most dramatic example of a slump occurs off the south coast of the island of Hawai'i. Here the Hilina slump is actively moving much of the south flank of Kīlauea Volcano in a southerly direction by as much as 4 inches per year. Periodically, rapid adjustments take place. Such an adjustment took place on November 29, 1975, and the 1868 earthquake had been a previous adjustment. It now appears that slumping and debris avalanches are part of the normal growth and aging process of volcanic islands. Could the birth of volcanic islands pose tsunami dangers as well?

Local Volcanic Tsunamis

Could Hawaiian volcanoes produce an explosion like Krakatau and generate a giant tsunami? Fortunately, this is extremely unlikely. Hawaiian-type volcanoes tend to be effusive and not explosive, and in historical times only one local tsunami has been related to volcanic activity. This tsunami occurred in 1919, during an eruption of Mauna Loa. Large volumes of lava had been pouring into the sea for several days when at 7:30 in the morning of October 2, Mr. Carlsmith of Hilo witnessed the tsunami while standing with his family on the Ho'ōpūloa wharf.

> My first intimation that a tidal wave was impending was the recession of the sea. Suddenly it seemed to slope backward from the shore. In a moment it was visibly running downhill. The rugged rocks of the coastline were exposed and stranded fish were left flopping on the shore. . . . Then the water came rushing back. I should say they [*sic*] were 12 to 14 feet higher than high water mark.

One of his sons ran toward their automobile to try to save it, but he was forced to abandon it as it was wrecked by the surging water. The other son was washed into the wharf shed, where he managed to grab a beam and hang on as the wave receded. Carlsmith himself was washed off the pier and carried about a hundred yards offshore. He struggled through the turbulence and managed to regain the land. Here he found his sons safe, but his wife was gone. She had been carried almost a quarter of a mile out to sea, where she could be seen struggling to remain afloat. Frantically, Carlsmith searched for a canoe, but all those on shore had been wrecked by the waves. Responding to his pleas for help, two fearless

Hawaiian men swam out to a canoe floating offshore and then paddled out to rescue Mrs. Carlsmith. Following the tsunami, the October 4 *Hilo Daily Tribune* reported: "It is now nearly impossible to get a Hawaiian boatman to take spectators to view the cascading of the lava into the sea by boat from Hoopuloa. Frankly, they admit fear of another tidal wave, although many believe the real reason is superstition regarding the wrath of Pele."

The exact cause of this tsunami is not definitely known, but the Hawaiian Volcano Observatory registered no unusual earthquake activity during the period. One possibility is that as the lava continued to flow into the sea, the front of the delta it built became increasingly steep until part of it collapsed, slumping down slope. This tsunami, then, though ultimately a result of volcanic activity, may really have been caused by a submarine landslide.

Landslides on active lava flows have been observed more recently both above and below water. On December 4, 1986, Dr. Lee Tepley, a scientist and professional photographer, was scuba diving in about 50 feet of water offshore of Kalapana on an active lava flow from Kīlauea Volcano. He and another professional photographer, Jim Watt, were observing liquid magma forming pillow-shaped masses of lava rock, when with no warning a small underwater explosion occurred. Watt immediately swam away from the lava slope. "I felt these vibrations next to the wall and bailed out, backpedalled real quick," Watt said. "This huge section collapsed like an avalanche." According to Watt, Tepley disappeared under a mass of volcanic rock. "He vanished. I was sure he was killed," Watt recalled in the December 6 *Honolulu Star-Bulletin*. Tepley didn't hear the explosion and was caught in the subsequent lava landslide and pulled down at least 200 feet. "Everything got black and surgy and suddenly I got pushed straight down in the middle of all these jagged rocks and boulders and debris," said Tepley. As an experienced diver, Tepley knew that he could die from the bends (decompression sickness) if he surfaced too suddenly, but he didn't have enough air to stay down. He quickly swam to the surface and, taking a new tank, returned to depth in order to decompress. Tepley was later treated at Hilo Hospital and released, vowing never to dive on an active lava flow again.

The most recent evidence that volcanism could produce a tsunami comes not from any of the island volcanoes, but from an active volcanic

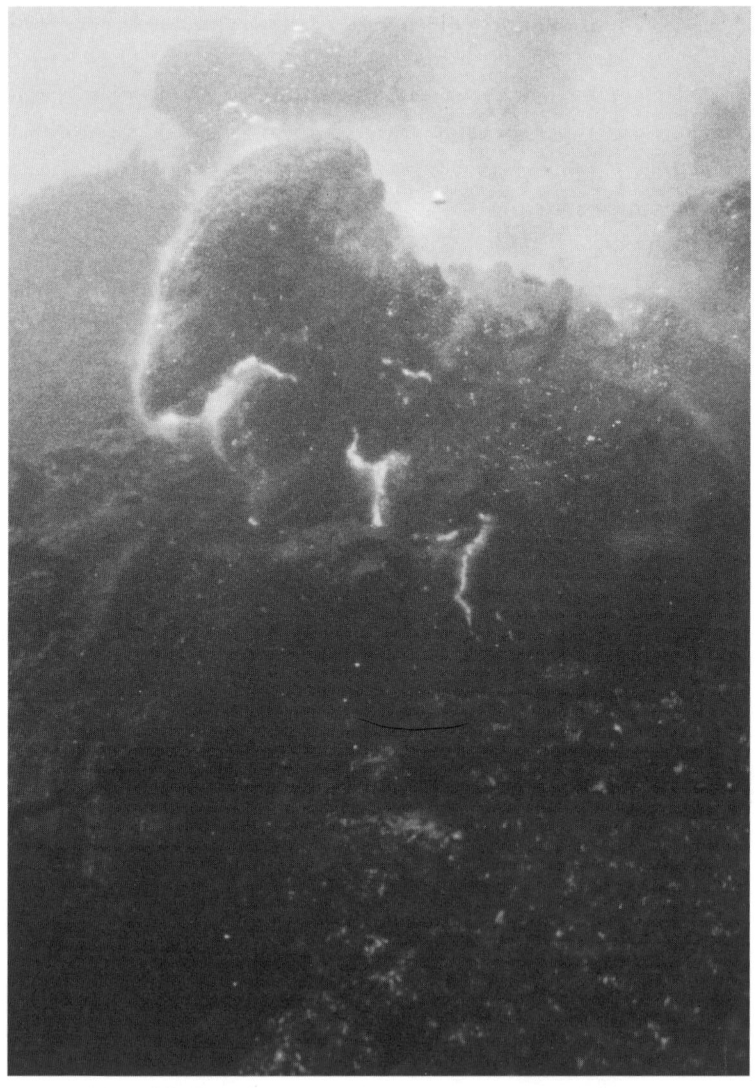

Figure 7.9 Pillow lava forming underwater on a steep landslide-prone slope on the flank of Kīlauea Volcano off Kalapana on the island of Hawai‘i.

seamount named Lō‘ihi. Hawaiian for "long one," Lō‘ihi is located about 15 miles southeast of the island of Hawai‘i. It has been known for some time that the volcano is growing toward the surface (and may someday become the newest Hawaiian island), but not until the summer of 1996 did scientists begin to have a taste of just how violent her growth pains

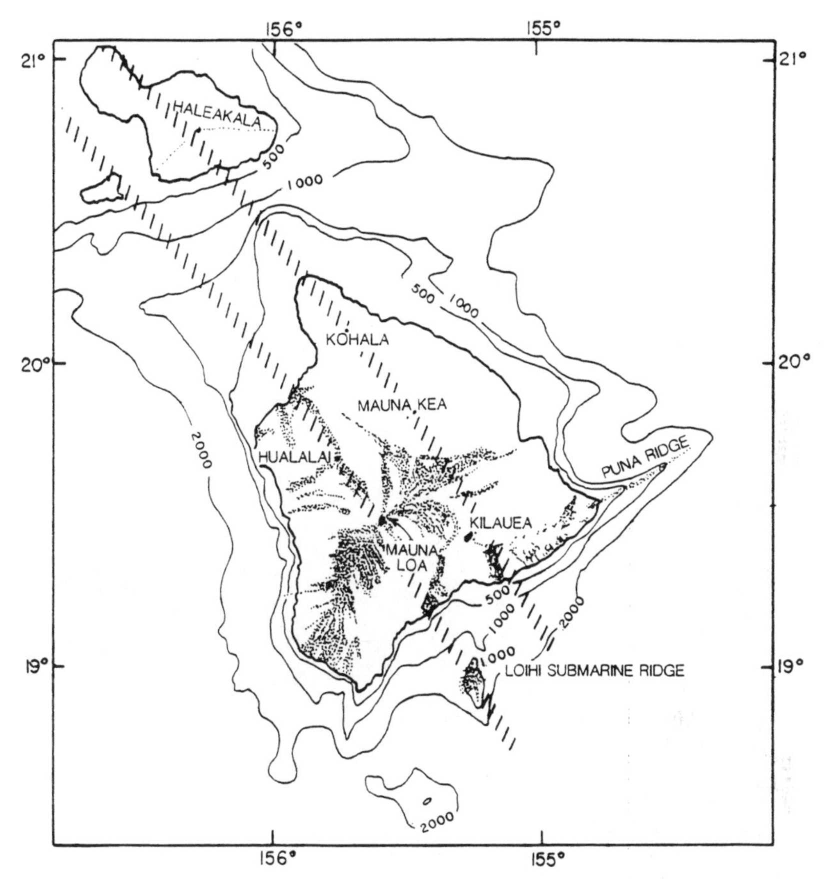

Figure 7.10 Map showing the location of Lō'ihi seamount off the south coast of the island of Hawai'i.

might become. On July 16, a major series of earthquakes began. Over the next three weeks, more than 4,000 tremors would be recorded by the Hawaiian Volcano Observatory, the largest swarm of earthquakes ever measured on any Hawaiian volcano. More than 40 of these would have magnitudes measured between 4 and 5 on the Richter scale. It was obvious that something big was up down on Lō'ihi. Research vessels were dispatched to Lō'ihi in August and September, and a submersible would dive to the summit to document the changes in the volcano.

One cruise deployed "sonobuoys," that is, underwater microphones on free-floating buoys, which radioed sounds from Lō'ihi back to the research vessel. The sounds included crackling and grinding noises interpreted as

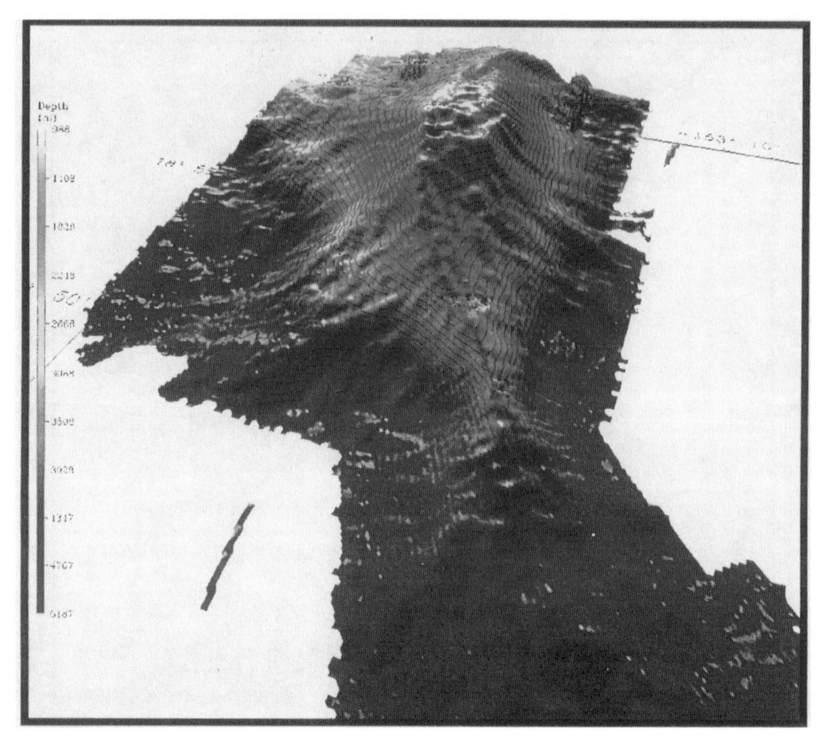

Figure 7.11 Three-dimensional image of Lō'ihi seamount created from multichannel echo sounding data.

indications of a major eruption in progress. A multichannel sonar survey was conducted to provide an updated map of the underwater topography of Lō'ihi; it revealed major changes in the shape of the summit in the area that had produced the earthquakes. Prior to the seismic activity, the shallowest area of the summit had been a zone of hot water vents rising to a depth of 3,215 feet. The new maps showed that this area had completely collapsed, forming a pit crater with its bottom nearly 1,000 feet deeper. Never before had scientist observed such an event, the actual formation of a pit crater on a mid-ocean, hot-spot volcano.

The Hawai'i Undersea Research Lab (HURL) submersible *Pisces* dove into the new pit crater. The scientists on board saw that most of Lō'ihi's surface was covered by large, broken boulders and pillow-shaped masses of hardened basaltic lava. Then the scientists found evidence of large landslides in several areas of the volcano. In the course of not more than 3 weeks, Lō'ihi had been transformed in a violent cataclysm. The summit had collapsed, a pit crater had formed, and large submarine landslides had occurred. "This was a Mount St. Helens–sized volcanic event," said Dr. Alex Malahoff, director of HURL. He went on to add that such an event could have triggered a destructive local tsunami throughout the Hawaiian Islands had it occurred quickly and not spread over several days. One more source of local tsunamis had surfaced.

The Threat of Local Tsunamis

What are the chances of another great 'Ālika slide tsunami or of a tsunami from volcanic activity on Lō'ihi seamount? No one knows for sure. And with large earthquakes occurring only about once a century, do residents of Hawai'i really need to be concerned about local tsunamis? The answer is an unqualified yes. Since 1848, there has been no period longer than 35 years without a local tsunami in the Hawaiian Islands.

Because of the very short interval between the occurrence of an earthquake in Hawai'i and the arrival of tsunami waves in populated areas, it may not always be possible to issue a tsunami warning in time. What can be done to avoid being caught in a locally generated tsunami? The Tsunami Warning System advises that people in coastal areas immediately evacuate anytime a strong earthquake strikes. But they should also remember that tsunamis produced by landslides or volcanic events may not be associated with strong earthquakes. And, of course, you cannot feel the earthquake producing a tsunami at a distant location. Therefore, if you notice a sudden drop or rise in sea level, this may be your only warning that a tsunami is in progress. Move to high ground immediately, or in the frank words of Hawai'i County Civil Defense Administrator Harry Kim, "Run like hell!"

U.S. DEPARTMENT OF COMMERCE
National Oceanic and Atmospheric Administration
National Weather Service

AND YOU MUST HOLD ONTO SOMETHING TO KEEP FROM FALLING...

HEAD FOR HIGH GROUND...

Figure 7.12 Tsunami warning poster indicating what to do when a strong earthquake occurs.

8

Recent Tsunamis around the World

Even though the Hawaiian Islands have not been assaulted by a major Pacific-wide tsunami for more than 30 years, tsunamis continue to be a real and present danger in the Pacific region. Hardly a year goes by without at least one destructive tsunami striking somewhere in the Pacific. Unfortunately, the press pays little attention to these disasters, occurring as they often do in remote countries. This lack of press coverage has led many people to falsely believe that little threat exists from tsunamis. The following accounts of some of the locally destructive tsunamis of the last two decades show that there are valuable lessons to be learned from each of these disasters.

Sea of Japan Tsunami, 1983

Just before 9 A.M. local time on May 26, 1983, a large earthquake measuring 7.7 on the Richter scale occurred in northern Japan. At 9:14 A.M. the Japanese issued a tsunami warning of the highest degree, "great tsunami," for coastal areas along the Sea of Japan in northern Honshu. But the first tsunami waves had already reached the coast of Aomori and Akita prefectures by 9:08 A.M. The tsunami would be called the Nihonkai-Chubu earthquake tsunami or the 1983 Sea of Japan tsunami. The death toll from the waves would eventually reach 103, most of whom were tourists or recreational fisherman. The death toll could have been much worse. The western coast of Japan does not have the V-shaped bays that have resulted in prodigious tsunami run-up along the eastern Sanriku

Figure 8.1 Seismogram of May 6, 1983, Japanese earthquake recording a magnitude of 7.7 at the Honolulu Observatory.

coast. Also, many of the towns and villages facing the Sea of Japan are built well above sea level to protect them from the high waves produced by the winter monsoon that blows across the Sea of Japan from Asia. Many towns along this coast are further shielded from the high winter waves by seawalls. At some of these towns the height of the tsunami was less than that of the seawalls, and there was little damage. For example, one town at the north end of the Oga Peninsula is protected by an 18-foot seawall. The tsunami surged up to about 15 feet here, and only one house was damaged. Fortunately the tsunami struck after the winter storm period was over, or the tsunami inundation might have added to already high winter seas.

But seawalls did not protect those who were down at the beach or in coastal harbors. In front of the seawall, conditions were very different. A school group of 45 children, with two teachers and two of the fathers, had been on an outing and were headed along the coastal highway some 30 feet above sea level. They had just taken an exit from the highway and were starting down a sloping road toward the beach when the earthquake struck. The tremors were not strongly felt on board the bus, and no one really thought about a tsunami. In fact, many people living along the coast of the Sea of Japan believed large tsunamis were a feature of the Pacific Ocean side, and never struck their coast.

The school group continued on to the beach, where they sat on rocks having a picnic lunch. Nearby a forty-seven-year-old fisherman watched as the sea withdrew slightly 15 minutes after the earthquake. Then the wave came back in as a mighty surge, picking up the school group and carrying them offshore with its withdrawal. They were washed into an area about 1,000 feet from the beach where the current had slackened. Here they floated around a small rocky island until local fishermen came to their rescue. Thirteen children drowned, but amazingly 32 of the children and all four adults were saved.

On the other side of the Oga Peninsula sits the seaside Oga Aquarium, a popular tourist attraction. That day there were about 200 visitors either touring the aquarium or wandering around outside enjoying the sea breeze and ocean view. Inside the aquarium, one of the employees, Mr. T. Hosoi, had felt the earthquake. Because he had previously experienced a tsunami elsewhere in Japan, he began quietly watching the area offshore of the aquarium. Only 7 minutes after the earthquake, he saw unusual wave activity in the distance that he correctly interpreted as the oncoming wave front of the approaching tsunami. Mr. Hosoi immediately made a warning announcement over the public address system, telling the aquarium visitors to evacuate the building and head to higher ground. He then left the aquarium with a portable electric megaphone and began warning tourists outside the building. One of the visitors outside the aquarium that day was Mr. S. Sato, who after moving to higher ground took the remarkable series of photographs shown in Figure 8.2. The first picture is taken 13 minutes after the earthquake and shows the water rapidly rising and overflowing a retaining wall. The photos that follow were taken at 15- to 20-second intervals and show the water surging up to its maximum inundation before rapidly withdrawing in what one observer described as a mighty waterfall.

Mr. Hosoi's knowledge of tsunamis and quick thinking may have saved the lives of as many as 200 tourists at the aquarium. There was, however, one needless tragedy. A Swiss couple was standing outside the aquarium when Hosoi made his warning announcements. Being from Switzerland, they probably knew little about the threat of tsunami waves and may not

Figure 8.2 May 1983 Sea of Japan tsunami. Series of three photographs taken by Mr. S. Sato in front of Oga Aquarium during the tsunami.

have understood his warning, for they did not flee the oncoming waves in time and were swept out to sea. The husband somehow managed to reach safety, but his wife was lost to the tsunami.

Another tragedy occurred at the construction site of the new Noshiro Thermal Power Plant at Noshiro Harbor. There were 306 workers at the busy coastal building site, where large revetments were being used to enclose massive concrete caissons. Some 67 vessels of various sizes were employed on the project as well as a number of deep-sea divers. The first warning was a tsunami alert heard over the radio by one of the engineers, Mr. K. Yaguchi. He quickly went up on to the roof of a building, where he could see the tsunami already surging into the construction site. At about the same time, the radio warning was heard in the construction company office, where they immediately transmitted a message to workers with hand-held radios down in the caissons. But it was already too late—the tsunami waves were picking up vessels and surging into the caissons. The tsunami death toll at the construction site would reach 34.

The loss of life on vessels was restricted primarily to workers on relatively small boats. Ship captains said that vessels more than twice as long as the height of the waves were able to ride out the tsunami by turning the bows of their vessels into the oncoming waves. Smaller boats, however, were flipped end over end (pitch-poled) by the oncoming waves and destroyed. Nine lives were lost in the boats.

In the caissons, the rubble rock used for construction became lethal as it was picked up and tossed about inside the revetments. A total of 25 were killed inside the caissons, including nine deep-sea divers. There had been no way to alert the divers, let alone save them.

The run-up at Noshiro Harbor was over 25 feet, but just 8 miles north of Noshiro the tsunami surged in to a height of 45 feet. Waves continued to crash ashore throughout the night and well into the next day as the tsunami was repeatedly reflected back and forth across the Sea of Japan. Even two days later, small tsunami waves could be recognized at some tide stations.

The fishing industry suffered heavy loss, with more than 250 boats sunk outright and some 1,600 vessels washed away or damaged. Crop fields were flooded and many houses destroyed, resulting in damage totaling $800 million.

Scientists try to learn something from every tsunami disaster. They, of course, want to better understand the phenomenon and improve their predictive capabilities, but they also hope that lives will not have been lost in vain—that through what they learn fewer innocent lives will be taken by the next tsunami. It was known that as tsunami waves approach shore, they sometimes form bores upon entering shallow water, as happened at Hilo in 1960. But prior to the 1983 Sea of Japan tsunami, many researchers felt that as a tsunami wave broke and formed a bore, much of the energy was dissipated, thereby producing smaller run-up heights and hence less damage along the shore. The bores that formed along the western coast of the Akita Prefecture struck with tremendous force. At beaches armored with 4-ton concrete tetrapods, the structures were picked up and scattered like children's jacks, some carried over 400 feet inland. Researchers now theorize that "very strong turbulence accumulates at the front of a bore as it approaches shore releasing the energy along the shoreline." Bores were also observed traveling up several rivers, in some cases as breaking waves and in others as a series of gentle, undulating swells.

There were other costly lessons learned from the Sea of Japan tsunami. Prior to 1983 the official policy of the Japanese Meteorological Agency (JMA), which has the responsibility for tsunami warnings in Japan, was to issue a tsunami warning within 20 minutes following an earthquake. This 20-minute time lag was largely a holdover from the days when computers were big, expensive, and rare in Japan. By 1983 computers had become readily available, and 20 minutes was no longer really necessary to provide a warning. In spite of this, JMA did not issue its official warning until 20 minutes after the earthquake. By this time tsunami waves were already striking most coastal areas. Many of the tsunami deaths could have been prevented had there been even a few minutes of additional warning time.

The case of the Swiss couple at the Oga Aquarium points to the need to make special considerations for tourists. It is very probable that tourists from areas where tsunamis are not common will not be aware of the danger. Also, any tsunami warning messages delivered in areas frequented by tourists should be announced in at least one additional language other than Japanese. This should be a language such as English, which many tourists are likely to understand.

The tragic deaths at the Noshiro Harbor construction site point not only to the need for more advance warning of an approaching tsunami, but also to the need for a tsunami evacuation plan and some means to alert divers working underwater at coastal construction sites.

In an interesting observation made after this tsunami, a Japanese scientist compared victims struck by a tsunami wave with the famous cliff divers of Acapulco. Since the divers strike the water at speeds greater than tsunami waves strike their victims, why is the tsunami death toll so high? To paraphrase the scientist, "the tsunami victims hit something solid"—like a diver at Acapulco plunging onto the rocks.

Chile Tsunami, 1985

On March 3, 1985, a major earthquake again struck Chile. Measuring 7.4 on the Richter scale, the epicenter lay off the coast near Valparaíso. The earthquake claimed 124 lives and more than 2,000 people were injured in Valparaíso and Santiago, Chile. A small tsunami was generated, which measured less than 4 feet at Valparaíso but advanced across the Pacific and was recorded at many tide stations of the Tsunami Warning System. At Papeete, Tahiti, the tsunami waves measured 4 inches and at Honolulu just over 1 inch. But Hilo, always sensitive to tsunamis, would record waves nearly 20 inches high, second only to Valparaíso.

Mexico Tsunami, 1985

The great Mexican earthquake of 1985 also generated a small tsunami. At 6:18 A.M. on September 19, the main earthquake, measuring 8.1 on the Richter scale, struck Mexico. Centered about 250 miles southwest of Mexico City, the quake had a devastating effect on the heavily populated metropolis, resulting in a death toll mounting to 10,000 and leaving over 600,000 homeless. The tsunami generated by this earthquake spread from its source across the Pacific and down the coast of Mexico. At the town of Lazaro Gardenas, closest to the epicenter, the largest waves were about 9 feet high and ran ashore as far as 180 feet inland. Along the Mexican coast from Manzanillo to Acapulco the tsunami waves ranged from 3 feet to about 10 feet high. At Hilo the waves measured 9 inches and at

Tahiti the tsunami was only 2 inches high. How is it that such a giant earthquake generated such a small tsunami? Some scientists believe that the collision of the tectonic plates off the Pacific coast of Mexico produces little vertical movement of the sea floor, hence little displacement of the sea water above the epicenter of the earthquake.

Aleutians, 1986

On May 7, 1986, at 2247 GMT, an earthquake measuring 7.6 on the Richter scale struck the Aleutian Islands. The epicenter was located about 100 miles southeast of Adak Island and approximately 70 miles southwest of Atka Village, Alaska. Less than 30 minutes later, the Pacific Tsunami Warning Center had sent out messages to tide stations requesting confirmation of a tsunami. The Adak tide gauge station, where a wave nearly 6 feet high was registered, confirmed the generation of a tsunami. A regional tsunami warning was issued by the Alaska Tsunami Warning Center at 2315 GMT and a Pacific-wide tsunami warning was issued at 2351 GMT (1:51 P.M. Hawai'i time) by the Pacific Tsunami Warning Center.

The epicenter of this earthquake was very close to that of the March 9, 1957, 8.3 magnitude earthquake, which had generated a Pacific-wide destructive tsunami (see Chapter 4). A report came in from the Midway tide station, which recorded waves over 2 feet high, similar wave heights to those registered at Midway during the 1957 tsunami. Concern grew as the waves headed for the main Hawaiian islands.

The earthquake had, indeed, generated a Pacific-wide tsunami that would be recorded at tide stations throughout the Pacific. But fortunately, a large destructive tsunami failed to materialize. At 0510 GMT the Pacific Tsunami Warning Center canceled the warning, with predictions that maximum sea level fluctuations of 50 centimeters (about 20 inches) would be measured at some tide stations. Honolulu would actually record 40 centimeters, Hilo 55 centimeters, Hanalei, Kaua'i, 91 centimeters, and Kapa'a, Kaua'i, up to 122 centimeters, but no damage would result.

Both the Alaska Tsunami Warning Center and the Pacific Tsunami Warning Center had done an excellent job in providing a timely tsunami warning to the Pacific basin. The tsunami warning had been issued only 64 minutes after the earthquake itself.

Though the warning system had operated smoothly, the Civil Defense alert and evacuation procedures had not gone as well. The Emergency Broadcast System had failed to work; incredible as it seems, there had been no regular testing procedures. It seems that a "streamlined emergency communication system" set up just a year before had failed when the Hawaiian Telephone hookup malfunctioned.

Evacuation efforts, however, did result in thousands of beachgoers, tourists, and workers in coastal areas moving inland. Hundreds of yachtsmen took their boats to sea, where they could be seen offshore of Waikīkī. Police tried to keep the evacuation moving smoothly, but by 4 P.M. most of the main streets in Honolulu had become hopelessly jammed. The normal afternoon rush hour got an early start with the closing of state and county offices and private businesses, and thousands of parents rushing to pick up their children from school. According to a Civil Defense spokesmen, "Kalanianaole Highway was a parking lot. The Moanalua Freeway was a parking lot." In a lame response to criticisms concerning the traffic problems created by simultaneously closing government offices and schools, Civil Defense stated that they had followed the O'ahu evacuation plan, but "we don't have enough roads" to handle the traffic.

Even in quiet Hilo traffic was snarled. And just as in 1960, there were Hilo residents who refused to believe there was anything to fear; it was "tsunami party time!" On O'ahu it was reported that a solid line of traffic was actually headed *toward* Waikīkī just before the first wave was due to strike.

Whereas businesses, schools, and government offices outside the evacuation zone had shut down nearly all at once, many stores and shops in the tsunami inundation zone stayed open. In some areas, residents simply refused to leave the seashore, continuing to swim, surf, and fish.

Once again, radio disc jockeys added to the confusion with unofficial and inaccurate reports of wave activity. The international press reported that there had been "no tsunami," and to many in Hawai'i the 1986 alert would become known as "Waveless Wednesday." Indeed, the alert was viewed by most people as a false alarm, and confidence in the warning system was further undermined. But the Tsunami Warning System had responded to the large earthquake in the only way possible, with prudence and caution. Far better that a warning be issued and people evacuated from threatened areas and have only small tsunami waves arrive, than be caught unaware by giant waves like those of 1946, which brought horrible death and enormous destruction.

There were many lessons to be learned from the 1986 experience. It was not just the public that didn't understand the threat; even police and other emergency personnel did not have a clear idea of what to expect. Perhaps the May 7, 1986, tsunami should best be viewed as an exercise, but a very expensive exercise. The estimated cost of the alert, including lost business activity, was put at $30 million. But following the experience, new evacuation zones were drawn up based on precise calculations of expected tsunami inundation. Many of these were smaller than the previous zones, reducing the number of people that needed to be evacuated. The emergency broadcast system was modified, and since 1986 the telephone company has had no involvement. Government agencies desperately needed to learn from the experience and be better prepared for the next tsunami.

But how much had we really learned? On the tenth anniversary of the May 1986 tsunami, the following quote was published in the monthly newsletter sent out by the Hawaiian Electric Company: "If Kansans know what to do in a tornado, and Floridians know what to do in a hurricane, why don't Hawai'i people know what to do in a tsunami?"

Nicaragua, 1992

Since the devastating tsunamis of the early 1960s, scientists had made considerable progress in understanding earthquakes. In 1973 seismologists originated the concept of a "seismic gap." A seismic gap is a normally active segment of the boundary of a tectonic plate that has not been ruptured by a large earthquake for 30 years or more. Such gaps are considered likely locations for future earthquakes. In fact, nearly all large earthquakes in the Pacific have occurred within seismic gaps. In 1981 three American scientists with the U.S. Geological Survey and one from the Institute of Seismic Investigations in Nicaragua published a paper in the prestigious journal *Science*. In this article they identified a seismic gap off the coast of Nicaragua that had not had a large earthquake since 1898. They went on to show that similar areas off the coast of Mexico and Central America had large earthquakes on the average every 50 years, suggesting that "a large earthquake in this area [off Nicaragua] might be long overdue," and that it might have a Richter magnitude as large as 7.5, similar to the 1898 quake.

Eleven years would go by before the scientists' prophetic words would be proven true. At about 8 P.M. on the evening of September 2, 1992, a moderate-sized quake struck off the Pacific coast of Nicaragua. Little information was available from the local seismic network, as it had fallen into disrepair due to the Nicaraguan Civil War, but seismographs in the United States, Europe, and Japan would record the event and assign the earthquake a Richter magnitude of 7.2. Only about half the coastal residents even felt the quake, and they described it as "weak" and "soft." But the displacement of the fault may have been as much as 10 times greater than suggested by the size of the earthquake, and there was considerable motion on the sea floor. An area 120 miles long by 60 miles wide slowly rose about a foot, producing an enormous "bump" of water, which then surged ashore as a single destructive tsunami wave. The wave struck the coast just an hour after high tide, producing run-up of over 30 feet at the village of El Transito, and rising over 20 feet high in many areas along the coast. The tsunami trapped mostly children and elderly adults who

Figure 8.3 September 1992 tsunami damage at El Transito, Nicaragua. This area, where 16 people were killed (14 children and 2 elderly men) was the most devastated by the tsunami. Nearly all the houses in El Transito, some 200, were destroyed by the tsunami, which reached more than 30 feet high here.

were caught in their sleep or unable to flee from the wave. Many healthy adults apparently managed to outrun the tsunami and survived. Casualties amounted to 167 dead and 500 injured. Destruction was widespread with over 1,500 houses demolished, leaving an estimated 13,000 people homeless.

Though the tsunami was a major disaster along the coast of Nicaragua where the energy was concentrated, the tsunami lost power as it spread out across the Pacific. The wave would measure about 3 feet in the Galapagos and at Easter Island, and only 4 inches in Hilo.

Flores Island, Indonesia, 1992

The islands that make up the Indonesia Archipelago span a zone where three huge tectonic plates collide. The Indo-Australia plate, the Asian plate, and the Pacific plate all converge and thrust against one another, producing some of the largest and most frequent quakes on earth. About 1,100 miles east of Jakarta, just to the northwest of Timor, lies Flores Island. At 1:30 P.M. on December 12, 1992, a 7.8 Richter magnitude earthquake rocked the ocean floor off the north coast of Flores. The sea floor was suddenly elevated 4 feet, raising the entire column of water above it and creating a deadly tsunami. Within minutes the waves were crashing ashore on the eastern end of Flores Island.

The village of Wuring, built on a low sand spit only 5 feet high and 2,000 feet long, was extremely vulnerable. The tsunami surged ashore 10 feet high at Wuring—high enough to destroy 80 percent of the wooden houses in the village and kill 87 people. The damage may have been made worse by the fact that the village fishing fleet was moored off the front of the village. The waves picked up the boats and washed them into the houses. The concrete mosque was one of the few buildings to survive.

The destruction was even worse on the far eastern end of the north coast of Flores Island. Here the waves averaged over 33 feet high, while elsewhere they attained no more than 23 feet. Near Cape Bunga, the northeasternmost point of the island, the sea surged in to the incredible height of over 85 feet at the village of Riang-Kroko, and it averaged 65 feet at Leworahang, running inland for a distance of nearly 2,000 feet. Riang-Kroko was completely obliterated. Of more than 200 houses in the village,

Figure 8.4 December 1992 tsunami at Wuhring, Flores Island, Indonesia.

Figure 8.5 December 1992 tsunami at Riang-Kroko, Indonesia. The run-up here was 86 feet high and washed nearly 2,000 feet inland, removing all traces of human habitation.

not a single dwelling was left. Even the house foundations were totally removed, as well as nearly every coconut tree and most other vegetation in the village. Out of a total village population of 406, the tsunami killed 137 people. According to a survey crew that studied the site following the tsunami, all that was left was "brown, bare ground, scattered with large chunks of debris and coral."

Not only were the tsunami waves largest on the eastern end of Flores, but they surged ashore sooner than expected. At the port of Larantuka, the tsunami arrived only 2 minutes after the earthquake, much too soon to have traveled from the region of the epicenter. Why did these waves arrive so quickly, and why were they so much larger than those striking regions farther west? Scientists speculate that the waves that struck the eastern end of Flores Island were not from the main tectonic tsunami, but from a secondary tsunami generated by a large submarine landslide set off by the earthquake.

Another interesting phenomenon, with a tragic consequence, was observed at Babi Island. This tiny island, less than 1½ miles in diameter, is located just 3 miles off the north shore of Flores Island. The north side of Babi Island faces the Flores Sea and is bordered by a wide coral reef. Toward the protected south side of Babi Island, facing Flores, the reef becomes much narrower, and in the middle lies a small tidal flat where two fishing villages had been built. At the eastern side of the flat was the Christian village of Pagaraman, and to the west the Moslem village of Kampungbaru. The site would seem an ideal location, protected from ocean swells and even waves generated by local winds. But the location did not protect the villages from tsunami waves. Unlike ocean swells and local wind-generated waves, whose energy is concentrated near the surface, tsunami waves, with their energy spread throughout the full depth of the water, can penetrate into even sheltered coastal areas without a significant loss of energy.

Only 3 minutes after the earthquake the tsunami approached Babi. As the wave surged toward the tiny island, it was split in two, one wave advancing around the east side, the other traveling around the west. These waves were probably largest right at the shoreline and had their crests moving perpendicular to the shore. The two waves wrapping around the island met on the "protected" south side, just opposite the two villages. The collision of the two waves greatly amplified their height, and some

Figure 8.6 December 1992 tsunami at Pagaraman on Babi Island, Indonesia.

scientists believe that the tsunami may have been further amplified as part of the wave energy was bounced off the coast of Flores Island and returned as a reflected wave. The resulting tsunami surged across the tidal flat into the two villages with great destructive power. The run-up at the Christian village of Pagaraman exceeded 18 feet and was just over 15 feet in the Moslem village of Kampungbaru. Both villages were totally destroyed. There was not a trace of a building at the Christian village, and in the Moslem village even the mosque was carried away, leaving only the roof lying on barren ground. Out of a population of 1093, a total of 263 people were killed by the tsunami. The area laid waste presented a grisly scene. Observers described seeing human remains hanging from the branches of the few surviving trees in the village. Many residents had tried to flee during the earthquake, only to have their path blocked by rocks falling down from the high ground behind the villages.

On and around Flores Island the tsunami had taken a tremendous toll. Nearly a quarter of the population of Babi Island was killed and almost one-third had died at Riang-Kroko. In all nearly 1,000 people had been killed by the tsunami, the death toll heaviest among the very young and very old. There had been no evacuation plans and most residents had no information about tsunamis. The natural response when people saw the water surging in had been to run. Children and old people didn't run fast enough and were caught by the waves.

Another trend emerged from the death statistics on Babi Island. Though the gender ratio in the population was approximately equal, almost twice as many women (175) were killed as men (88). A similar trend has been observed in other tsunami disasters. Scientists studying the 1992 Flores Island tsunami suggested that men and women had different motivations driving their response to the crisis. The men they felt "tend to take refuge for themselves," whereas the women "want to protect small children and the elderly during an emergency," often paying for their altruism with their lives.

1993 Hokkaido (Nansei-Oki) Tsunami

At 10:17 P.M. on July 12, 1993, a large earthquake struck off the west coast of Hokkaido, Japan. The magnitude measured 7.8 on the Richter scale and the epicenter was located just 50 miles offshore of the island of Okushiri in the Sea of Japan. Within less than 5 minutes computers at JMA had calculated the areas of greatest tsunami danger and warnings were issued. There was no 20-minute delay for calculation as there had been during the 1983 Sea of Japan tsunami, but there were other delays. This time the problem lay in getting the warning sent to the threatened areas. The Hokkaido Prefecture branch office in Hiyama received the tsunami warning by radio, but because their transmitter was "busy with other business," the warning was sent by fax over ordinary telephone lines, losing 5 valuable minutes. When telephone lines on Hokkaido became overcrowded, the telephone system began to automatically cut connections. Worse yet, Okushiri town did not receive the fax at all because the telephone lines to Okushiri Island had been put out of service by the earthquake; no message was received at Setane town because the fax machine could not operate due to a power outage caused by the earthquake; and the town of Aonae on the south end of Okushiri Island became isolated from telephone communication when their fiber optic cable was damaged by the earthquake.

But even if the warning had been received just 5 minutes after the earthquake, it would have been too late for much of the coast of Okushiri Island. Here, waves ranging in height from 10 to 35 feet were already crashing ashore.

Monai village, located on the western side of Okushiri Island directly opposite the area of generation of the tsunami, was struck by the waves almost immediately after the earthquake and totally destroyed. The village was inundated to a height of 65 feet, destroying all 12 houses and killing 10 people. Just north of Monai, the tsunami surged into a narrow valley, washing up to a height of over 100 feet, one of the highest run-ups ever recorded in Japan.

Even at the less exposed north and south ends of Okushiri Island, the tsunami caused tremendous destruction. The shoaling bathymetry at both ends of the island caused the tsunami waves to be refracted and wrap around the island. At the northern end of the island at Inaho village the tsunami came ashore with waves over 33 feet high, killing 13 residents and destroying every house in the village. But the greatest single toll of death and destruction would be at the south end of Okushiri, at the town of Aonae.

Aonae, with its population of 1,600, sat "securely" behind its seawalls. The seawalls, built following the 1983 tsunami, were constructed to prevent inundation by tsunami waves 15 feet high—the height of the 1983 tsunami. But only 4 to 5 minutes after the 1993 earthquake, tsunami waves over 30 feet high struck the town, easily overflowing the seawall. The water attained speeds in excess of 40 miles per hour as it surged through the town, leaving the southern end of Aonae completely devastated. The only surviving structure was a reinforced concrete public toilet; all 82 wooden houses were carried away. Even the northern part of Aonae was not spared. Here a 25-foot-high sand dune was covered with a pine forest planted as a wind break. The tsunami passed over the dune and through the forest, leveled more than 1,000 trees, and swept away about 100 houses. In many areas the tsunami did not go over the seawall but invaded through passageways—one such passageway may have been created during the tsunami by a large fishing boat being slammed against the seawall.

Just 7 minutes after the first wave struck, the second wave surged into the fishing harbor from the east, sinking boats at their moorings and carrying others into the town. Two boats caught on fire, probably from electric short circuits at their batteries. Cars were also affected by short circuits, and many submerged vehicles had bright headlights on and horns madly sounding. Burning boats were carried into the main part of town, where supplies of kerosene and propane used for cooking quickly caught fire and spread the flames. In all, 340 homes not devastated by the tsunami were destroyed by the fires. Fire trucks were unable to reach the burning sections of town because the streets were filled with debris left by the tsunami waves. A total of 107 people were killed by the tsunami at Aonae, and an additional two died from the fires.

Even the sheltered east side of Okushiri Island suffered tsunami damage. The village of Hamatsuma, northeast of Aonae, had run-up of over 65 feet that destroyed 36 wooden houses and caused 32 deaths. The high run-up here may have been produced as wave energy was focused by local submarine topography, or perhaps similar to the phenomenon observed at Babi Island, by a combination of reflected waves and refracted waves meeting at Hamatsuma.

The tsunami also struck the large island of Hokkaido. The area di-

Figure 8.7 July 1993 tsunami at Okushiri Island, Japan. A view of tsunami and fire damage at Aonae on the southeast end of Okushiri Island.

Figure 8.8 July 1993 tsunami at Okushiri Island, Japan. Squid fishing boat washed into Aonae. To the left is a damaged fire truck.

rectly opposite the epicenter of the earthquake was the hardest hit by the tsunami, with wave run-up heights from 15 to 30 feet. The approximate times at which the waves struck were determined from electric clocks, which had shorted out due to submersion in sea water. Clocks were found at eight different locations, including offices, homes, and wrecked automobiles. On some parts of Hokkaido (near Setane and Ota), the tsunami arrived too early, within only 3 to 5 minutes after the earthquake. It has been suggested that these tsunami waves may have come from a different source area, in a way similar to the way tsunami waves struck the far eastern end of Flores Island in 1992.

The effects of the 1993 Hokkaido tsunami were also felt outside Japan. Only 30 minutes after the earthquake, waves up to 13 feet high struck Russia, causing over $6 million in damage. South Korea was hit 90 minutes after the earthquake with run-up of 6 feet but little damage. All of the fatalities and most of the destruction were concentrated at Okushiri Island and along the west coast of Hokkaido. The death toll from the 1993 Hokkaido tsunami reached 202 with 28 missing and 83 severely injured. More than 1,000 homes were seriously damaged or totally destroyed

Figure 8.9 July 1993 tsunami at Okushiri Island, Japan. Battery-operated clock stopped at the time of inundation by the tsunami, some 15 minutes after the earthquake.

and 1,729 vessels were lost or damaged, totaling some $600 million in property losses.

There were several lessons learned from the 1993 Hokkaido tsunami. It wasn't sufficient for the JMA to declare a warning if the message could not be delivered to those areas in jeopardy. Tsunami warnings must be sent out over secure, high-priority communication channels, links that will not be destroyed by the very earthquakes that may generate tsunamis. But even with rapid communication, there may be some areas too close to the source of the tsunami to have ample warning time. For such areas, tsunami education is a must. Some of the residents of Aonae remembered the 1983 Sea of Japan earthquake and tsunami. When the violent shaking started, they got up and began to evacuate. Some climbed steps leading to higher ground behind the village as the fastest way to evacuate and this quick thinking saved their lives. Others tried to evacuate with their cars and were stuck in traffic jams and overwhelmed by the waves. Still others became casualties when they returned to retrieve valuables. Finally, many died at Aonae while waiting for family members or neighbors to evacuate. There is an old saying, a kind of rule, in the Sanriku district, that region of Japan most frequently struck by tsunamis. It is said that this rule is handed down from generation to generation in order to reduce casualties. It goes, "Tsunami ten-den-ko," and means, "In a tsunami, every man for himself."

East Java, 1994

On June 2, 1994, at 1818 GMT, an earthquake measuring 7.2 on the Richter scale struck the region of the Java Trench in the northeast Indian Ocean. It was 1:18 A.M. local time (on June 3) in nearby Indonesia, and though the epicenter of the quake was located only about 120 miles from the coasts of Java and Bali, less than 20 percent of the islanders felt the tremors—most just continued sleeping. Earthquakes are a part of everyday life in this part of Indonesia and no one thought that this one might be extraordinary. Like the earthquake off Nicaragua two years earlier, there was no strong ground shaking and no earthquake damage was reported. But the quake lasted for over 2½ minutes, during which time the pent-up seismic energy was gradually released.

About 50 minutes after the earthquake, tsunami waves as high as 36 feet struck the coast of Java. Survivors said that there was no warning. According to the *Jakarta Post*, they heard a roaring sound at the last minute, which they first thought was "the sound of the wind" until they were suddenly hit by water.

Three villages along the coast of Java were particularly hard hit: Pancer, Lampon, and Rajekwesi. Pancer village was situated on a sandbar lying between a river channel and the Indian Ocean. Such a location meant that the village fishing fleet could be moored in the protected waters of the river during storms. But such a site was also extremely vulnerable to destruction from tsunami waves. Not only could tsunami waves wash completely over the sand bar on which the village was built, but the villagers were sandwiched between the river and the ocean with little in the way of a secure escape route. The tsunami came ashore with waves 30 feet high at Pancer and penetrated more than 2,500 feet inland. In all 126 people were killed and 671 homes and other buildings damaged or totally destroyed. But there were two mitigating factors. A dense row of palm trees planted at the front of the village may have helped reduce the force of the incoming surge, and there was apparently no powerful withdrawal of water ("rundown"), so the river channel may have "contained the advancing water front."

About 4 miles west of Pancer lay Rajekwesi village. Situated on a small pocket beach, Rajekwesi suffered the worst damage of any village as a result of the tsunami. At the east end of the village the tsunami surged up nearly 46 feet above sea level, killing 33 people. The beach was completely washed away, and the village was literally flattened over a distance of some 1,300 feet. About 100 homes or other buildings were destroyed, including nearly all homes along the beach.

Lampon village, like Pancer, was built on a sand dune at the mouth of a small river. Here the tsunami apparently traveled into the river and then invaded the village from both the ocean and river side; there was nowhere to run. The wave surged up over 30 feet, killing 40 villagers and destroying 144 buildings. The foundations were washed away and even the ground itself was deeply eroded. Along the coast of Java the toll of casualties would eventually reach 223 dead and 440 seriously injured. There were 1,226 buildings badly damaged or destroyed and 940 vessels damaged.

Though death and destruction were concentrated in the three villages, the effects of the tsunami were felt elsewhere. The tsunami struck two world-renowned surfing spots in Java. P. T. Plengkung Indah Wisata and so-called G-camp are both meccas for traveling surfers. Fortunately the tsunami struck in the middle of the night or there might have been high casualties. Eyewitnesses described the coral reefs fronting the camps as being "ripped apart" at many sites. A 65-foot section of reef was reportedly "shaved off" by more than 3 feet and large chunks of reef debris were observed on the beach following the tsunami.

It was at Sampu Island, an uninhabited nature preserve, where one of the most curious effects of the tsunami was observed. On the side of the island facing away from the origin of the tsunami, a long, narrow depression was eroded into a forested valley for over 150 feet, cutting inland from the beach. Scientists hypothesize that this "highly concentrated erosion" was produced as two waves advanced around the island from opposite sides met and splashed onto the shore. This interpretation is supported by accounts of fishermen, who observed one wave approaching from the southwest and another approaching from the northeast around the island.

The tsunami also struck the northwest coast of Australia, roughly 1,000 miles from Java, 3 to 4 hours after the earthquake. The tsunami surged through a gap in the coral reef and over the recreational beach and parking lot at Baudin near Northwest Cape. There waves estimated at over 12 feet carried fish, crayfish, rocks, and chunks of coral for a distance of nearly 1,000 feet inland. Fortunately the area was deserted at the early morning hour, but residents of a nearby trailer park described hearing a noise like the "roar of a train." At about 6:15 A.M. local time a liquid natural gas ship was tossed around by the surge and a minor oil spill was created as the hose transferring fuel between two ships was broken. There were scattered reports of inundation at other areas along the coast where gaps in the coral reef left passages for the tsunami to strike the exposed shore.

The lessons of this tsunami are not unlike those of others around the Pacific—lack of knowledge of tsunamis by the local population meant death to villagers along the coast of Java. Most people did not realize that even a moderate earthquake has the potential to generate a devastating local tsunami. Given half a chance, people will try to save themselves. At the village of Gerangan, in the western part of East Java, there was no loss of life, even though the village was struck by tsunami waves over 18 feet high. Here the residents fled to high ground following the second wave and did not return until the danger was over. It was the third wave at Gerangan that had the highest run-up.

The tsunami waves striking the coast of western Australia were the first in over a century, since the eruption of Krakatau in 1883. Just because tsunamis are infrequent does not mean that they pose no danger. Western Australia needs a contingency plan, a warning system, and tsunami education. Even in areas where tsunamis are rare, the population still needs to know how to respond to the deadly waves.

1994 Shikotan (Kuril Islands) Earthquake Tsunami

At 10:23 P.M., October 4, 1994 (1323 GMT, Oct. 3), the earth's crust 13 miles beneath the sea floor off Russia's southern Kuril Islands, northeast of Hokkaido, Japan, began to move. Seismic waves radiated across and through the earth, triggering seismic alarms in Japan at the JMA tsunami warning center and in Russia at the Sakhalin Tsunami Center. An earthquake measuring 8.1 on the Richter scale was occurring. On Shikotan Island, the island closest to the epicenter, the ground shaking became extremely violent. Some 16 people, mostly Russian military personnel, were killed as Soviet-era poured-concrete buildings, many constructed without steel reinforcing rods, crumbled to the ground. Meanwhile the sea floor in the area above the earthquake's epicenter rose by as much as 6 feet and caused a huge bulge of water to form, thus generating a tsunami.

It was 3:23 A.M. at the Pacific Tsunami Warning Center (PTWC) in Hawai'i. Here too the seismic waves had triggered the alarm and the two scientists on duty, Bruce Turner and Chip McCreery, rushed to the observatory and began quickly analyzing the earthquake data.

In Japan, with the painful lessons of the 1983 and 1993 tsunamis taken to heart, the JMA issued their tsunami warning after only a few minutes. By about 5 minutes after the earthquake, NHK (Japan Broadcasting Corporation) had already broadcast its first warning message with a map of the Pacific coast of Hokkaido flashing in red and that of the main island, Honshu, flashing in yellow. JMA had adopted an easy-to-understand color code for their televised warning, with red indicating a "tsunami warning" with predicted wave heights up to 2 meters (6.6 feet) and yellow for "tsunami caution" with waves to 1 meter (3.3 feet). Gratefully missing this time was the most lethal warning level, "great tsunami," with predicted wave heights of 3 meters or more (greater than 10 feet). Along with this visual display was a written warning in both Japanese and English instructing coastal residents of the threatened areas to evacuate to high ground. The use of English to inform foreign visitors corrected another weakness discovered following the 1983 Sea of Japan tsunami.

Within only a few more minutes NHK had begun to broadcast the first JMA predictions for tsunami arrival times along the coast of Japan. The first wave was due to hit the northernmost tip of Hokkaido at 11 P.M. local time, just 37 minutes after the earthquake.

But waves were already striking the southern Kuril Islands. At the town of Yuzhno-Kurilsk on Kunashire Island, the tsunami surged in to a level of 10 feet above sea level. Many houses in the oldest part of town facing the sea were damaged and two bridges were destroyed. Two 300-ton fishing vessels were washed ashore and another sunk. The earthquake and tsunami were the strongest to hit the southern Kurils in the twentieth century, but though 16 people died as a result of the earthquake, there were no casualties directly related to the tsunami. The main reason for this is the high level of public awareness of the tsunami threat. There have been at least 27 tsunamis to strike the southern Kurils in historical times and the population was aware of the danger and knew how to respond.

As if on cue, the first tsunami waves to strike Japan hit Nemuro, at the northern tip of Hokkaido, at 10:58 P.M., 2 minutes ahead of schedule. Here the water came ashore to a height of 1.73 meters (5¾ feet), just under the maximum 2-meter prediction. But then things took a turn for the worse. A little more than an hour after the earthquake, JMA increased the warning level for Honshu to a full tsunami warning. During a tsunami warning, coastal areas are to be completely evacuated, but during the lower level tsunami caution, certain officials are instructed to go to the coast and observe changes in the level of the water. Some of these observers were so busy watching for sea level changes that they did not hear the upgraded warning. If a large tsunami had struck, they might well have been among the first casualties.

Meanwhile at the PTWC, the small staff had been inundated with telephone calls from Alaska, California, Oregon, Washington, and as far away as Chile and Australia. The only vacant telephone line was being used alternately to send telexes and to receive computer data from tide stations. This revealed a serious weakness at the international center operated by the United States to serve 26 different countries in the Pacific. Finally at 4:23 A.M. local time in Hawai'i, one hour after the earthquake, with confirmation of tsunami waves striking Japan, the PTWC issued a Pacific-wide tsunami warning. The warning included coastal regions of Hawai'i, and the west coasts of the United States and Canada.

Civil Defense officials in Hawai'i, already aware of the earthquake, quickly met and decided that in order to avoid the traffic gridlock experienced during the 1986 tsunami evacuation, they would sound the alarms an hour earlier than called for in the manual. At such an early morning hour most people would still be at home, not yet on the road, so the streets would be free for the necessary evacuation of residents from threatened low-lying areas. The state school superintendent was contacted and at 6:20 A.M. the decision made to close all 238 public schools throughout the state of Hawai'i. All nine campuses of the University of Hawai'i were also to be closed and all government workers were advised to stay home until the all clear was announced. At 6:30 A.M. the coastal sirens were sounded and evacuation shelters opened.

I (W.D.) had been awakened myself by an early morning telephone call and quickly prepared to drive into Hilo, pick up camera equipment, and take up my post as a tsunami observer on the roof of the Naniloa Hotel overlooking Hilo Bay. As I drove into town, the tsunami was crossing the north Pacific Ocean. At 7:19 A.M. (Hawaiian time) the tsunami had reached Wake Island, registering a mere 6½ inches on the tide gauge. But after the 1960 experience with 6-inch waves at Christmas Island becoming 36-foot monsters in Hilo Bay, no one was yet ready to discount the possibility of a disaster.

With camera equipment loaded in my vehicle, I now drove toward Banyan Drive, the small peninsula projecting into Hilo Bay, where all of the resort hotels in Hilo are located. The highest building, the Bayshore Towers condominium, sits on the opposite shore of the bay, and tsunami specialist George Curtis had already set up his observation post on the roof there. I soon approached a roadblock manned by Hawaii County police. After I had shown my special Civil Defense pass and explained my mission, an officer removed a portion of the barricade and waved me through, shaking his head in disbelief as I drove into the inundation zone.

It was now nearly 8 A.M. and the tsunami had reached Midway Island, registering 20 inches on the tide gauge. This reading was similar to that of the 1957 tsunami, when 32-foot waves had struck Kaua'i. Bill Mass, a geophysicist at PTWC, commented in *The Honolulu Advertiser* (October 5, 1994): "that's bigger than I'd hoped. We're fairly concerned." It had been nearly 35 years since the disastrous tsunami of 1960; was this going to be the next big one?

At 8:02 A.M. the tsunami reached Guam, where it registered only 7½ inches. Was this tsunami to be a devastating natural disaster or a "false alarm," a public relations disaster for the warning system and Civil Defense?

I had now reached the Naniloa Hotel and entering the lobby was met by the maintenance manager. Personnel were busy loading the contents of the gift shops into vans, while other workers took care of important records. One of the employees asked me if I was going to leave my car parked in front of the hotel. "Why?" I responded. "Well, it may not be there when this is over." Then it struck me, all of downtown Hilo could be laid waste in the next few hours!

After setting up video and time-lapse movie cameras on the roof (with my colleague John Coney) and establishing radio contact with George Curtis on the other side of the bay, we waited. After a while, Bruce Turner at PTWC called us on our cellular telephone with an update, expressing his concern about the 20-inch wave at Midway. It was now 9 o'clock and the tsunami was due to strike Kaua'i at 10:28 A.M., O'ahu at 10:42 A.M., and Hilo at 11:08 A.M.

At 9:30 A.M. the Honolulu police set up roadblocks barring access to all roads leading to the coast on O'ahu. Some roads were completely closed, including the main artery of Ala Moana Boulevard. But a special "tsunami bus service" was put into operation with six different routes around the island. These buses, bearing special tsunami evacuation signs, took people from low-lying areas to higher ground. All beaches and beach parks were closed and the Polynesian Cultural Center shut down for the alert.

Meanwhile evacuation procedures, planned and unplanned, were taking place throughout the islands. Rather than attempting to evacuate the tens of thousands of visitors from Waikīkī, a policy of "vertical evacuation" was adopted for this densely populated tourist area. In vertical evacuation, people in solid, reinforced concrete buildings are moved above the floors likely to be inundated by the waves. The Moana Hotel offered a free buffet in the third floor Surfrider Tower ballroom to some 300 guests evacuated from the low-rise wings of the hotel. At Waikīkī's largest hotel, the Hilton Hawaiian Village, guests staying in rooms below the third floor were contacted by hotel management and given the opportunity to move to a higher floor. Meanwhile, workers filled pillowcases with sand, creating emergency sand bags in order to protect the shops and restaurants on the ground floor of the hotel. Works of art, Persian carpets, parrots, and penguins were all evacuated. The penguins were taken to a shelter on the second floor of the parking garage, while the parrots waited out the tsunami on an upper floor of the hotel. Thanks to good prior planning, the hotel had leaflets printed in both English and Japanese explaining the tsunami danger and the need for the evacuation.

State Civil Defense had also thought of the need for foreign language warnings and had prepared tapes for radio broadcast. However, none of these were used during the alert. The explanation from a Civil Defense spokesman was: "Everything was happening fast and furious. . . . We just didn't have time" (*The Honolulu Advertiser*, October 5, 1994). But the need was definitely there. The article in the *Advertiser* continued: "Foreign language speakers were left confused or clueless to the potential danger when safety officials used only English over loudspeakers to warn people." Police officers were observed shouting at Japanese tourists in English, "You are in danger! Please move to higher ground," while the tourists just stared at the officers in total incomprehension.

It was not just foreign speakers who didn't understand the message. One group of American tourists understood the alert to be a "Salami Warning" and wanted to know what all the fuss was about. Meanwhile, Mayor Jeremy Harris was busy conducting interviews with radio stations across the country to encourage their listeners not to cancel their Hawaiian vacation plans. More than once he had to explain to a distant disc jockey that tsunamis are different from hurricanes.

On the island of Maui, most people heeded the warnings, but there were some who remained on the beaches. There was nothing Civil Defense or the police could do. As Maui Police Chief Howard Tagomori explained, "we don't have the authority to force anyone to evacuate" (*The Honolulu Advertiser*, October 5, 1994).

On the Big Island of Hawai'i, Civil Defense chief Harry Kim ordered all coastal areas evacuated by 9:30 A.M. On the Kona side of the island at the Hilton Waikaloa resort, some 1,200 guests were transported from the beachfront hotel to the King's Course golf club further inland. Netting was carefully spread around the dolphin enclosure at the resort to protect the captive dolphins from injury if waves surged into their lagoon. In downtown Hilo, people could be seen double-parked in front of their businesses, hurriedly loading important records and computer equipment prior to evacuating the area.

From the roof of the Naniloa, I could see small boats heading out of Hilo Bay to form an irregular flotilla bobbing about a mile offshore, a safe distance from the dangerous shoal waters near the coast. The same exodus was occurring on O'ahu, where sailboats and speed launches of every description headed out of the Ala Wai Boat Harbor and Ke'ehi Lagoon. Even the Atlantis tourist submarine moved offshore to safe waters. In Honolulu Harbor large container ships began casting off and moving out to sea, shepherded by Coast Guard cutters and surrounded by commercial fishing boats. Off the island of Maui, the interisland cruise ship *Constitution* remained at sea rather than come into port at Kahului as planned.

Perhaps the most unusual evacuation of the morning was taking place 550 miles northwest of Honolulu. Four women and three men were on remote Tern Island studying endangered Hawaiian monk seals, sea turtles, and migratory birds in the Northwestern Hawaiian Islands Wildlife Refuge. A twin-engine prop plane, which stops once a month on the island, just happened to be there on the morning of the alert. The plane could take only five of the seven, so the wildlife refuge manager and assistant manager stayed behind. They climbed on to the roof of the generator building, barely 20 feet high, and hoped for the best.

Meanwhile up on the roof of the Naniloa my cellular telephone rang with a request from a local radio station for an on-the-spot interview. I tried to stress the real danger of the situation, desperately hoping that nothing was really going to happen.

Elsewhere others had already decided that nothing was going to happen. Off one of the renowned surfing beaches of the north shore of O'ahu, a dozen surfers could be seen enjoying the first major swell of the winter surf season. At 8:30 A.M. a police helicopter had hovered above the surf for several minutes while an officer announced, "Surfers below, a tidal wave warning has been issued. We request you leave the water and move to higher ground" (*The Honolulu Advertiser,* October 5, 1994). He was ignored. An additional surfer paddled out, bringing the number to 13. One long-time surfer overlooking the scene is quoted in the *Advertiser* as saying he "thought for a moment that it was an unlucky number" but continued that such "was superstition and based on ignorance." Not aware of the irony of his statement, he continued to sit observing people who pride themselves on their knowledge of the waves, waiting for what would be practically certain death should destructive tsunami waves arrive. The situation was much the same on Maui, where at 9:30 A.M. lifeguards on jet skis had ordered surfers out of the water, only to be ignored.

With the predicted time of arrival only moments away, a man stood on the breakwater in Waikīkī casting for fish. Another man ran across Kalākaua Avenue, down Waikīkī Beach and dove into the ocean. Then at the very moment the first tsunami wave was due to strike Honolulu, a woman with dark flowing hair, wearing a purple muumuu, crossed the street, and calmly marched into the waves. She gazed toward the horizon while reverently touching the water and then motioned toward her heart (*The Honolulu Advertiser,* October 5, 1994).

From our perch above Hilo Bay, we anxiously waited as the times of arrival at each island passed. Was the coast of Kaua'i being ravaged? Was Waikīkī a disaster area, the harbor at Kahului destroyed? Our radio and cell phone were strangely silent. Had the warning center itself been inundated? It was nearly dead calm and the water in Hilo Bay had a mirrorlike sheen. Then at 10:50 A.M. the water took on a strange, dull look. A muddy plume surged out of the Wailoa River into the bay, and a series of gigantic ripples began moving back and forth across the water. Was this the first wave or perhaps a precursor to a giant withdrawal before a devastating wave surged in? Twenty minutes went past and the predicted arrived time for Hilo came, but no large wave appeared. Perhaps the second or third waves would come ashore higher. No one knew. Soon a single sailing yacht left the fleet offshore and headed into the bay. As it approached the Reeds Bay mooring area, the County helicopter swooped down in front of the vessel and warned the skipper that the alert was not over. But the warning was simply ignored and the sailboat continued in to the shallow water of Reeds Bay. If a significant tsunami wave had struck at that moment, the vessel would have been dashed against the rocks along the shore. We trained one of our cameras on the boat, but this time luck was with her.

Now my radio came to life with reports of wave arrivals. The tsunami had reached Nāwiliwili Harbor, Kaua'i, at 10:20 A.M. with a run-up of 14 inches. Hale'iwa, O'ahu, had been hit by a 22-inch wave at 10:30 A.M., and Kahului, Maui, struck at 10:44 A.M. by a wave nearly 3 feet high. In Hilo Bay the water had surged to a height of less than 1½ feet. Finally at 11:55 A.M., when no destructive waves had been reported outside the

Figure 8.10 Sailboat ignoring warnings from Hawai'i County helicopter at the mouth of Reeds Bay during the tsunami warning of October 4, 1994.

Kuril Island area, the warning was canceled. By this time, more than 400 surfers had already paddled out to catch the early winter swell.

Officials greeted the cancellation with a mixture of relief, embarrassment, and concern for the credibility of the warning system. They seemed to be apologizing to the public for the inconvenience of the alert, but there had really been no choice. To quote Bill Mass of PTWC, "If we make errors, it should be on the side of safety. . . . People might have been inconvenienced, but nobody got killed." Roy Price, vice director of State Civil Defense, stated their view: "I can assure you the tsunami did arrive as scheduled. But it was a foot and a half high, not over two feet high. And thank goodness" (*The Honolulu Advertiser*, October 5, 1994). Mayor Harris is quoted as saying, "Well, it looks like we dodged another bullet," and was quick to add, "My experience is in a job like this you always get criticized. But I'd much rather be criticized for conducting an evacuation and not having the wave arrive than having to explain the loss of life had the wave arrived and had we not planned for it." The mayor was clearly bracing for the coming criticism from the business community over the monetary loss and inconvenience of the "false alarm." As might have been expected following the event, a headline in *The Honolulu Advertiser* read "Detection still inexact science" and was followed by "Two false alarms in eight years" as the opening to an article on the reliability of the warning system.

The warning had been expensive. The Pearl Harbor Naval Shipyard had secured its four dry docks, evacuating employees. The estimated cost in lost productivity was $250,000. Two Hawai'i tour companies alone estimated their loss at $500,000. They practically had to shut down their entire operation. The same day's newspaper quotes the executive vice president of Roberts Tours Hawai'i: "You can't run many tours in Hawai'i that are not in tsunami-inundation areas." The total cost would ultimately rise into the millions. Had there been any return on the investment? What had we learned from the "exercise"?

The decision to begin the evacuation earlier than called for by official procedures had been wise—the traffic congestion of 1986 was avoided. Mayor Harris agreed that it was better to inconvenience some people earlier rather than take the chance of gridlock, confusion, and threat to life later. But other procedures had not worked so well. It clearly states in the telephone directory to stay off telephone lines during emergencies, to turn on your radio and listen for instructions. But in an informal poll of 20 people queried by a newspaper reporter, nearly all learned of the tsunami warning not from the sirens or the news but from worried relatives calling to warn them. This heavy use of the phone lines resulted in saturation of the system at times during the alert. Even worse, many people called 911 as if it were a tsunami information line—jamming the 911 emergency service.

Many people may simply not have heard sirens because some weren't working. Five of the warning sirens on Maui were not operating, and three on Kaua'i didn't work. In fact, the three sirens on Kaua'i had been broken since Hurricane 'Iniki, more than two years before. It seems that repair of the sirens is a county responsibility, but replacement is a state problem. However, the state claimed that sirens weren't repaired or replaced because they were awaiting Federal emergency funds. At Kapolei on O'ahu, there were no sirens at all—they had yet to be installed. Officials remarked that they should get one sometime in the next "two years."

The use by the Hilton Hawaiian Village of information pamphlets printed in both English and Japanese was a success. In fact, the Japanese tourists seemed to be the least worried about the tsunami. According to Japanese restaurant manager Kazuo Oki, "They know what a tsunami is, but they have never seen (one) for themselves" (*The Honolulu Advertiser,* October 5, 1994). His comments make an important point—fear of scaring tourists away from Hawai'i is largely unfounded. Japanese visitors learn about tsunamis in Japan and they were not particularly worried about them in Hawai'i. They understood the threat and knew the proper action to take.

The careful planning, staff training, and bilingual leaflets helped make the evacuation at the Hilton among the smoothest anywhere in Hawai'i. "These are things we learned from hurricane Iniki," said the hotel's general manager in an interview with *The Honolulu Advertiser*. It is a sad commentary on the state of tsunami preparedness in Hawai'i that we have to learn what to do for a tsunami alert from a hurricane disaster. Hawai'i, and the entire Pacific for that matter, needs tsunami preparedness and education programs. After all, tsunamis have killed more people in Hawai'i than hurricanes, earthquakes, and volcanic eruptions combined.

There were also lessons to be learned outside the Hawaiian Islands. The tsunami waves from the Kuril Islands had continued across the Pacific, surging in over 3½ feet high in Crescent City, California, demonstrating once again the vulnerability of this coastal town to tsunami waves. Finally in far-off Chile, the tsunami still had enough energy to register over 2½ feet on the tide gauge at Talcahuano, over 10,000 miles away.

Following most recent tsunamis, an international survey team has been dispatched to the affected areas in order to study the effects of the tsunami. Japanese scientists, among the world's greatest experts on tsunamis, have usually played a prominent role in these survey teams, but this time they were excluded. Japanese university scientists had been invited by the Russians to participate, but they were prohibited by their own government from taking part. They were told that the trip to the Kurils would be "illegal" and they "could lose their jobs." This stems from a 50-year-old dispute between Russia and Japan over the political status of the Kuril Islands. Japan has repeatedly sought the return of the southern Kurils, which were seized by the Soviets during the closing days of World War II. This political friction has held up Japanese aid to Russia, and Japan currently prohibits government officials, including university scientists, from traveling to the Kuril Islands. In spite of the lack of Japanese experts, the international survey team did a commendable job and made many useful observations.

On the side of Shikotan Island opposite the area of generation of the tsunami, waves were the largest, as might be expected. Here at least four tsunami waves came ashore and run-up averaged over 20 feet, whereas on the side of the island in the shadow zone, the tsunami run-up averaged only about 8 feet. However near Petrova Point on Kunashiri Island, seemingly protected behind Shikotan Island, anomalously high run-up values were recorded. Apparently tsunami waves traveled around both sides of Shikotan and met near Petrova Point, producing run-up of over 17 feet. This is similar to the situation at Babi Island, Indonesia, in 1992, where the superposition of two waves produced locally high run-up.

Another interesting finding had to do with sediments transported by tsunami waves. Surveys of the 1992 Nicaraguan tsunami, the 1992 Flores Island tsunami, and the 1993 Hokkaido tsunami all found extensive transport of sediment, beach rocks, and coral onto the shore. At Shikotan, however, there was essentially no transport of beach materials and even the vegetation that had been inundated by the tsunami waves was in good condition. This finding suggests that the wave run-up at Shikotan was a gradual process, consistent with the fact that run-up heights were fairly uniform regardless of differences in local topography. One explanation for this is that the initial tsunami wave was positive, not a recession of the water but an initial inundation. According to some scientists, initial positive waves produce "smaller run-up and less violent run-up motions" than an initial negative wave followed by a positive wave.[1] Tide records

[1]H. Yeh, V. Titov, V. Gusiakov, E. Pelinovsky, V. Khramushin, and V. Kaistrenko, The 1994 Shikotan earthquake tsunamis, *PAGEOPH* 144 (1995):855–874.

on both Shikotan and Hokkaido show the initial tsunami formation as a positive wave.

Study of the earthquake itself produced a troubling finding. Many large earthquakes have occurred in this general region due to subduction of the Pacific tectonic plate into the Kuril Trench. The 1994 earthquake occurred in the same region as a previous 1969 earthquake that measured 8.2 and generated tsunami waves with run-up of more than 16 feet. Calculations indicated that it should take 78 years for enough stress to accumulate for another such earthquake, yet only 25 years had passed. Was this a direct contradiction to the seismic gap hypothesis? Further study of the 1994 earthquake indicated that it was not relieving the stress of the tectonic plates being thrust beneath the Kuril Trench, but instead resulted from buckling of the plate. This meant that it was an earthquake within a tectonic plate and not between two plates. The seismic gap hypothesis generally applies only to the stresses that build up between plates.

There were several final ironies related to the 1994 Shikotan earthquake tsunami. The day of the tsunami warning in Hawai'i, October 4, was the date for the first planned meeting of the board of a proposed tsunami museum in Hilo. One of the principal goals of the museum was tsunami education. but the meeting had to be canceled due to the tsunami alert. Also on October 4, the scientist-in-charge of the PTWC in Hawai'i was in Washington, D.C., attending a meeting on the tsunami warning system and tsunami preparedness. And finally, it has been speculated that the devastation of Russian military and civilian facilities in the southern Kuril Islands might soften Russian resistance to their return to Japan, ultimately reuniting Shikotan, Kunashiri, and Iturup with the Japanese homeland to the south. This might free up much-needed economic aid for Russia. It is, indeed, an ill wind that blows no good.

1994 Skagway, Alaska, Tsunami

On November 3, 1994, almost exactly a month after the Shikotan event, another tsunami occurred in the north Pacific Ocean. This one was to be quite unusual; it would have no Pacific-wide consequences yet prove to be a killer.

On the eastern side of the harbor at Skagway, Alaska, work was being done to renovate the railroad dock. The dock was piled high with rock and soil fill material while an area between the dock and railway embankment was being filled in. Further adding to the weight on the dock was pile-driving equipment, altogether producing a load of some 23,350 tons. In late October one of the workmen had noticed a crack in the ground near the shoreline. Concerned about the safety of the project, he had quit. Then on November 2, additional movement of the ground caused some of the braces used in the construction to break.

At 7:10 P.M. on November 4, when the tide had dropped to 4 feet below normal low tide, the sediment supporting the heavy dock suddenly gave way and slid into deeper water. The old dock with everything on it, including two D7 Caterpillar bulldozers, was carried away. The slide was estimated to have been some 4,500 feet long, 600 feet wide, and 50 to 60 feet thick.

At the time of the slide, Paul Wallen of Homer, Alaska, along with his brother were working inside a steel caisson, when suddenly debris began pouring down on top of them. Paul's brother managed to climb out, but Paul was trapped inside. A 20- to 25-foot wave surged across the harbor, smashing into the Ferry Terminal on the opposite shore. The water ran up to an estimated height of over 36 feet above sea level along the shoreline. A floating concrete dock broke loose, smashing into the Broadway dock, breaking the passenger ramp off the dock, and sinking the auto ramp. A 4-inch fuel line snapped, sending 60 to 100 gallons of fuel oil onto the water.

Part of the tsunami wave poured into the small boat harbor, wreaking havoc as boats alternately struck bottom and smashed into each other. Two men on one of the boats saw the wave coming and began running down the floating dock toward shore. Unable to stand once the dock began bucking, they crawled to safety, but only after one man had injured his back and the other aggravated a heart condition.

Meanwhile a large return wave had smashed into the area of the railroad dock, killing Paul Wallen. His body would not be found for nearly three months. Damage was extensive. The harbor had been scoured from an original depth of 40 feet to over 90 feet in places. The small boat harbor suffered about $100,000 in damage, and the cost of repairs to the

Ferry Terminal were estimated at some $2,000,000. But the railroad dock was most severely affected, with replacement estimated at between $15 and $20 million.

No earthquake had triggered the slide—this was a non-seismic tsunami. Though such events are relatively rare, they are a serious hazard: there is no natural warning and no ground shaking to alert people to flee. The tsunami was noteworthy in another way—it had produced the first tsunami casualty in the United States since the tragedy at Halapē in 1975.

1995 Antofagasta, Chile

At 1:11 A.M. Sunday (0511 GMT), July 30, 1995, an earthquake measuring 7.6 on the Richter scale struck off the coast of northern Chile near the city of Antofagasta. The epicenter was only about 200 miles south of where the 1877 earthquake, which produced a large destructive Pacific-wide tsunami, had struck. Only 2 minutes after the earthquake, tsunami waves struck Caleta Blanco surging in up to 8 feet, and within 4 minutes the wave had reached Antofagasta, where the peak to trough height (amplitude) was over 9 feet.

Eyewitnesses reported that the first sign of the tsunami was a withdrawal of the water, but tide records indicate otherwise, showing the first movement as a rise in the water level. An initial small rise in water level is often overlooked by casual observers, whereas the subsequent withdrawal, usually much larger, often alerts onlookers by creating a noise as shells and pebbles are dragged seaward. The rise of sea level as the first motion of the tsunami indicates that the earthquake motion resulted in an uplift of the sea floor. The uplift could have been produced as the oceanic tectonic plate (Nazca plate) thrust under the continental plate (South American plate) beneath the Peru-Chile Trench, pushing up the sea floor on the continental side of the trench. In fact, an area onshore near the Mejillones Peninsula was uplifted, rising about 16 inches. This earthquake occurred in an area that had not experienced a seismic disturbance of this magnitude in historical times; hence the earthquake may partially fill a previously identified seismic gap.

The research vessel *Purihalar* of the Universidad de Antofagasta happened to be anchored in Caleta Bay during the tsunami. Her captain, Carlos Gurerra Correa, reported that his GPS[2] navigation instrument indicated that the ship was being washed around in "figure eights," while the ship's echo sounder showed the depth changing from 5 feet to over 35 feet as the tsunami waves surged in and out.

Following the earthquake, the PTWC had declared a tsunami warning for the area around Antofagasta and a tsunami watch for parts of the southeast Pacific. The warning was ultimately canceled when only small waves occurred outside the immediate area of the earthquake. But tsunami waves were traveling across the Pacific Ocean.

In the Marquesas Islands of French Polynesia, more than 4,000 miles from Chile, the tsunami came ashore as a 10-foot wave at Tahauku Bay on Hiva Oa. Two small boats were sunk, and an area of about 16 acres was inundated. The tsunami registered on the tide gauge at Hilo, where it measured about 2½ feet, and was recorded in Australia, Samoa, Mexico, California, Alaska, and Japan, though with smaller run-up values.

There were fortunately no casualties as a result of this tsunami, but once again luck played a major role. The tsunami struck nearly 3 hours after the peak of high tide, thus flooding was much less than it could have been. The timing was also important for another reason. Striking when it did, in the early morning hours on a winter's night, it arrived at a time when most small boats and fishing vessels were securely moored, the coastal resorts were empty, and no one was on the beaches. Had the tsunami struck during the middle of a summer day, casualties could have been quite extensive.

Once again the need for public education about tsunamis was brought home. The short time interval between the earthquake and the arrival of tsunami waves (as little as 2 minutes) would not have been sufficient for even the fastest warning system to operate. The only mitigating measure to save lives in such a situation is to educate the public to evacuate coastal areas at the onset of an earthquake.

---

[2]"Global Positioning System" is a highly accurate satellite-based worldwide navigation system.

Tsunamis in 1996

The years 1996–1998 were not without their tsunami disasters. Although none would be Pacific-wide in its destruction, many would be killers. Indonesia was struck again, not once but twice. On January 1, 1996, at 4:05 P.M. local time (0805 GMT) an earthquake measuring 7.8 on the Richter scale struck near the Makassar straits. While most people were still celebrating the New Year's holiday, tsunami waves inundated more than 70 miles of the coast of western central Sulawesi. Run-up measured over 16 feet and surged inland as far as a 1,000 feet. The effects were most severe at Tongglolbibi village where 9 people were killed, 63 injured, and 163 buildings destroyed.

Only six weeks later on February 17, a still larger earthquake, measuring 8.1 on the Richter scale, struck north of the island of New Guinea off the Indonesia province of Irian Jaya. A tsunami with waves up to 23 feet high came ashore, killing 107 people on Biak Island. At Yapen Island a total of 54 were reported dead or missing, and 10,000 were left homeless throughout Irian Jaya as a result of the tsunami.

Four days later on February 21, 1996, an earthquake measuring 6.6 on the Richter scale occurred on the opposite side of the Pacific, off the coast of northern Peru. A tsunami with run-up to heights greater than 16 feet came ashore at the city of Chimbote. There were 12 fatalities. Half those killed were line fishermen who were fishing off coastal rocks. Four people were gathering wood for fuel near the mouth of the Rio Santa when they were overwhelmed by the waves, and two men were panning for gold on the beach at Campo Santa when the tsunami surged ashore.

Papua New Guinea, 1998

The year 1997 and the first half of 1998 were relatively quiet, but then on July 17, 1998, a tsunami disaster once again struck the island of New Guinea, this time affecting the country of Papua New Guinea on the eastern half of the island. Papua New Guinea is a nation of some 4 million people, where many of the tribes still live a subsistence-based agricultural lifestyle. The El Niño of 1997/1998 had produced a serious drought in the region. Crops suffered, famine threatened, and the drying up of streams that are often the only source of water for drinking, cooking, and washing, resulted in serious health problems. Just as the country was beginning to recover from the El Niño, a disastrous tsunami crashed ashore along its northern coast.

At 6:49 P.M. local time (0849 GMT), an earthquake registering 7.1 on the Richter Scale rocked the ocean floor at the western end of the Bismark Sea, just offshore of the village of Aitape in West Sepik province, one of the most remote and isolated parts of New Guinea, and some 370 miles northwest of Port Moresby. Villagers reported their homes shaking from the tremors, but because most structures were made of materials gathered from the jungle, especially palm fronds, there was little in the way of damage. Nonetheless, the villagers were left with a sense of foreboding. Although this is a seismically active region of the Pacific, the local area had experienced only relatively moderate earthquakes in recent history, with but nine major earthquakes (magnitude greater than 7.5) occurring since 1900. A magnitude 7 earthquake had struck the area in December 1907, causing a stretch of coastline about 25 kilometers west of Aitape to subside, creating a large, shallow lagoon between 6 and 13 feet deep, which was named Sissano Lagoon. The same area was again struck in 1935 by a 7.9 earthquake, which may have generated a small tsunami. Though there appears to be no native lore of either of these events, the 1998 tsunami will long be remembered as one of the greatest tragedies ever to strike the region.

Within minutes after the earthquake, the Pacific Tsunami Warning Center had issued the following bulletin:

> THIS IS A TSUNAMI INFORMATION MESSAGE, NO ACTION REQUIRED. AN EARTHQUAKE, PRELIMINARY MAGNITUDE 7.1, OCCURRED AT 0850 17 JULY 1998, LOCATED NEAR LATITUDE 2S LONGITUDE 142E IN THE VICINITY OF NORTH OF NEW GUINEA.
> EVALUATION: NO DESTRUCTIVE PACIFIC-WIDE TSUNAMI THREAT EXISTS.

Yet with its shallow focus at a depth of between only 10 and 20 miles, the earthquake proved very efficient at generating a deadly local tsunami. Beginning shortly after 7 P.M. (0900 GMT) the first of three tsunami waves began to crash ashore, inundating a 20-mile stretch of beach extending from west of Aitape to the village of Serai. The worst damage was at the four villages of Sissano, Warapu, Arop, and Malol. All four villages were built on beaches, and their wooden huts were only a few feet above sea level. Warapu and Arop were especially vulnerable, situated between the Pacific Ocean and Sissano Lagoon on a narrow spit of land only 100 yards wide.

Many villagers spoke of hearing sounds like those made by jet fighters, just before the turbulent water surged through their villages. Some residents were crushed in their huts, others buried under sand and debris, while most may have drowned. At Warapu and Arop nothing was left standing, both villages completely washed away into either Sissano Lagoon or the Pacific Ocean.

The timing of the event added to the tragedy. Most of the village children were at home on a school holiday with their families, not staying as they normally did at the mission schools farther inland. The children were torn from their parents' arms as they all tried to flee the area. As in other tsunamis, the very young and very old numbered high among the casualties. They were too slow to run from the waves and too weak to climb coconut trees before being overwhelmed by the tsunami. The highest waves have been estimated at between 25 and 33 feet, enough to completely inundate the coastal villages.

Early reports told of 70 deaths, but this estimate was quickly and frequently revised upward. As this book went to press, government sources had placed the official death toll at 1,600; but with 2,000 still missing, this number was expected to rise, and many people felt that the final death toll would never be accurately known. Initially there had been hope that some of the missing would be found, as many villagers had reportedly fled inland and were thought to be huddled in small groups deep in the jungle. But as time passed and more bodies were discovered, hopes began to fade.

Of the survivors, many suffered from severe injuries sustained when they were thrown against trees and debris or impaled on sharp mangrove limbs and projecting roots. Others suffered from pneumonia, having inhaled water and then spent hours floating in the sea. Local medical facilities were quickly overwhelmed. Government officials reported that more than 550 people were critically injured. The 120-bed hospital at Vanimo treated more than 600 patients, with many lying on makeshift beds on the grass outside the hospital while awaiting treatment. Australia and New Zealand dispatched teams of doctors and nurses, and a portable field hospital was airlifted into the area. Local airstrips had been flooded by the tsunami, so aid was flown into the airstrip at Vanimo about 60 miles away, near the Indonesia border. As medical personnel began to treat the injuries, they found many broken bones as well as head and chest injuries; almost all the injured had lacerations. Bacteria-laden coral sand had infected the wounds, and gangrene had begun to set in. By the middle of the week following the disaster, doctors had been forced to amputate limbs from 22 men, women, and children. One doctor described the injuries as resembling battlefield wounds, and medical personnel began to worry that many of the tsunami survivors might die from infection or tropical diseases. The search for the living continued as four highly-trained "sniffer" dogs and their handlers from the Florida Rescue and Response Center were flown in to search the areas surrounding Sissano, Warapu, Arop, and Malol.

News of the disaster spread around the world, and expressions of sympathy began to pour in. Pope John Paul II dedicated his Sunday prayers to the victims, and Queen Elizabeth II, who is head of the Commonwealth and consequently also queen of Papua New Guinea, sent a message of condolence and sympathy to the people of the Aitape area. Even Papua New Guinea rebel leader Joseph Kabui requested that the people be told of his "deep sympathy." Nearly 5,000 people had been left homeless by the disaster, and aid for the survivors began to roll in. Steamship companies pledged to deliver relief supplies free of charge. Rotary International sent container-loads of medical supplies and equipment, and tarpaulins, ropes, water containers, and blankets. Chevron Oil Company loaned two aircraft and two nurses to the relief efforts, and the Australian Surfers Sunrise Club donated 24 rough-terrain wheelchairs for the amputees. Yet, confusion and lack of coordination threatened to disrupt relief efforts. Much of the food and supplies began piling up in warehouses in Aitape and Vanimo instead of being distributed. Part of the problem may have been that regional public servants, churches, and private aid organizations did not approve of the command structure set up under the state of emergency. There had simply not been adequate planning for a disaster of such magnitude.

As time passed, emergency rescue and evacuation efforts were replaced by searching for bodies and burying the dead. An estimated 500 bodies had been seen floating in Sissano Lagoon and the sea offshore, not counting the ones scattered through the mangrove swamps. Scavengers were drawn to the area to feed on the corpses, sharks and saltwater crocodiles in the water, pigs and wild dogs along the shore. Some 200 bodies were discovered more than 100 miles to the west in Jayapura, Indonesia, where residents avoided swimming in coastal waters or buying fish in the local markets.

As the search for bodies continued, it became impossible to identify the grossly disfigured victims, so authorities merely tried to keep count. Fear of disease began to take precedence over recovering the bodies, so rescue workers quickly either cremated the corpses with gasoline or buried them in shallow graves dug where the bodies were found. The Melanesian Council of Churches observed a special day of prayer for the victims, but this was a poor substitute for traditional burial ceremonies where family members spend time weeping over the bodies and telling stories. Papua New Guinea Defense Forces were planning to string barbed wire around the devastated areas in order to keep grieving survivors from returning to their villages to bury relatives and pick through the wreckage. As ponds, wells, and other freshwater sources became polluted, the risk of disease increased; government authorities made plans to evacuate survivors, relief workers, police, and military forces and then to seal off the area. Sissano Lagoon would be declared a mass graveyard. Colonel Kanene of the Papua New Guinea Defense Forces speculated that after the bodies had decomposed for a month or so, the narrow sandbar separating the lagoon from the sea might be blown up to allow the ocean to flush the remains out to sea and disinfect the area.

As with other recent tsunamis, plans were made to bring in a survey team of experts to see what might be learned from the disaster. Under the direction of Dr. Fumihiko Imamura of the Disaster Control Research Center at Tohoku University in Sendai, Japan, the team was to include seismologists, civil engineers, and tsunami experts from Japan, Australia, New Zealand, Russia, and the U. S. who would travel to the scene of the disaster after rescue operations had terminated. The scientists planned to study evidence such as sand deposits, stripped vegetation, debris on trees or wires, and water marks, as well as to record eyewitness accounts. They would produce maps showing the areas of inundation and try to determine what factors could have influenced this tsunami.

Almost from the beginning, some aspects of the tsunami resembled the local tsunamis in Nicaragua in 1992 and East Java in 1994. In each of these cases, a moderate-sized earthquake had generated a deadly local tsunami. What had caused such destructive wave run-up? Had a submarine landslide contributed to the tsunami? Or as Dr. Viacheslav Gusiaskov, head of the Tsunami Laboratory in Novosibirsk, Russia, speculated, one possible factor might have been "the presence of large depths (as much as 2½ miles deep) located close to the coastline." The survey would be important in answering many of these questions. But as the threat of disease grew, an on-site survey seemed less and less likely, and the scientists began to consider surveying the site by helicopters and taking photos and videos of the inundated areas.

As happens so often, there was an ironic twist to events in the aftermath of the tsunami. One of the donations to the stricken area was from the Stettin Bay Lumber Company, which gave lumber to rebuild local village schools destroyed during the tsunami. The tragedy is that there may be no need for new schools for many years to come, and even the mission schools that survived the tsunami may be closed. There are almost no children left to attend school. Nearly an entire generation was wiped out by the tsunami.

Even though no destructive Pacific-wide tsunamis occurred from 1992 through mid-1998, some 5,000 deaths resulted from local tsunamis. We learned from each of these disasters, and scientists were beginning to refine the science of tsunami prediction and to tackle the problem of false alarms. In the next chapter we will examine just what they have learned and what they predict for the future. We will also consider what we have yet to learn from the past.

9

How Much More Do We Know?

We have learned a great deal from the tsunamis of recent years. Even tsunamis of more than a century ago can still teach us about the phenomenon, when reviewed in the light of our current knowledge. Disasters that seemed particularly surprising and unexpected when they happened are now better understood, and this knowledge could help us prepare against future events.

The Sanriku Meiji Tsunami of 1896

In Japan, June 15 is celebrated as Boys' Festival. The festival in the year 1896 was a particularly happy one as many soldiers, just returned from the Sino-Japanese War, were being honored for their bravery. It was also the date when the annual banquet for the prefecture assembly was held in Kamaishi town, and in Utatsu village a wedding ceremony was taking place. It was a festival evening in almost every town and village along the rocky Sanriku coast of Japan. But only 93 miles offshore, an event was occurring that would make this particular festival infamous for years to come. Around 7:30 in the evening an earthquake began to rock the sea floor at a depth of about 6,000 feet on the western slope of the Japan Trench. Many people didn't even feel the tremors and those that did took no special notice in this land of frequent earthquakes—it was just another small quake, or so they thought. But unlike many of the earthquakes to which they had become accustomed, this one had little violent, rapid motion. Instead it held most of its energy in slower movement—movement that humans may only weakly perceive but that has the ability to generate powerful tsunami waves.

It was about 8 P.M., just after the fireworks had been fired at Ootsuchi town, that the tsunami struck the Sanriku coast. The first indication most villagers had was the sound, "a big noise." Then, according to one survivor, "Suddenly, water fell on me." In Kamaishi, everyone at the banquet, including all the town leaders, were overwhelmed and killed by the waves. At the wedding in Utatsu village, the bride, both families, and all the guests became victims of the tsunami. The groom alone would survive, to live out the rest of his days insane from grief and pitied by the other villagers.

The survivors were found by fishermen. Most of the fishing fleet had been offshore for the night fishing, as was the custom in this part of Japan, but four fishermen from Omoshige village had turned back toward shore. Through the darkness they began to see houses floating out from land and even thought they heard voices. Believing these to be apparitions,

Figure 9.1 A boy and his horse were caught in deep water during the 1896 tsunami. The boy survived by holding on to the horse's mane and tail while the animal swam to shore (painting on silk).

Figure 9.2 Prior to taking her bath a woman felt the 1896 earthquake, but because it was not strong began to bathe anyway. She was caught in her bath by the tsunami but survived. Her entire family was killed except for her two-year-old son (painting on silk).

perhaps the ghost ships that they had heard the older fishermen speak of, they remained silent, trembling in fear. The spell was finally broken when a loud voice called out, "I'm the deputy mayor, Yamazaki." The fishermen, startled out of their fear, began to help the poor survivors struggling among the wreckage.

A fisherman from Hirota village found an old woman floating on a piece of wood. When he helped her out of the water, he discovered that she was his own mother. But their happy reunion was short-lived. As soon as they reached shore, they realized that the rest of their family was gone: wife, children, brothers, and sisters, the entire village wiped out by the tsunami.

All along the Sanriku coast, the returning fishermen stared in horror. Said one, "When I came back, everything was gone. I didn't know what to do." The death tolls were truly staggering, the losses so great that instead of counting the victims, the authorities counted the survivors and subtracted from the former population totals. At Hirota village 126 out of

Figure 9.3 Karakuwa village. Next to the dead bodies of a small child and horse, a survivor stares blankly. In Karakuwa village, 248 houses out of 514 were washed away, and 836 people died.

371 people were killed; at Matsusaki, 272 out of 550 people killed; at Tarou village 1,867 out of 2,248 people killed; and at Touni village 766 out of 873 villagers died. All of the survivors had been at sea fishing, in the mountains grazing their horses and cattle, or out of the village for the festival—every last person present in the village had been killed by the tsunami. The death toll would eventually reach 28,321 and property damage totaled 6,222 houses destroyed. Run-up of 95 feet was recorded at Yoshihama, and the waves reportedly reached a terrifying 125 feet above sea level at Shirahama in Ryori Bay.

The gruesome task of recovering the bodies went on day after day. Most of the corpses were too damaged to be recognized, having been mutilated by collision with debris in the turbulent water. At Hirota village, the survivors used a net to recover bodies from their bay. On the first try the net caught over 50 bodies and was too heavy to pull in, so the villagers were forced to release it and pull in two smaller hauls. A pall of "sad smoking fires" hung for days over entire villages as the bodies were cremated.

A rescue effort was mounted to help the survivors, but it proved very difficult to bring relief supplies to many of the remote, isolated villages along the mountainous Sanriku coast. Little in the way of government aid was available as the national treasury had been exhausted by the war. Much of the affected area had not supported the ruling government during the Meiji Restoration, and furthermore the area was far away from Tokyo. Fortunately, private donations poured in from across Japan and from many foreign countries. The hardy people of the Sanriku coast rebuilt their villages, a little higher and a little farther from the sea.

Measuring Earthquakes

How was it that such a small earthquake, one barely felt, could have generated such a large tsunami? Similar cases where a seemingly small earthquake generated a large destructive tsunami occurred in 1992 in Nicaragua, where only about half the coastal residents described feeling the quake, and again in 1994 in Java and Bali, where less than 20 percent of the islanders noted any tremors. Earthquakes possessing this peculiar ability to generate tsunamis much larger than their Richter magnitude have been termed "tsunami earthquakes"; they have been the subject of intense study by seismologists since the late 1970s. Part of the answer to the mystery of tsunami earthquakes lay in how we actually measure the magnitude of earthquakes.

The best-known scale for reporting the strength of earthquakes was created in 1935 by Charles F. Richter of the California Institute of Technology. Richter developed his magnitude scale as a way to report the size of earthquakes occurring in southern California. It was really valid only for quakes recorded on certain seismograph instruments and covered only earthquakes in a range of about 400 miles from the seismograph. Richter's scale (M_L) was later revised and standardized for use worldwide. As presently defined, and presently called surface wave magnitude (M_S), it is based on the Raleigh wave, a type of seismic wave that travels along the earth's surface and is recorded on seismographs that measure only those surface waves with a frequency of around 20 seconds. Earthquakes actually generate seismic waves at many different frequencies, but surface waves with a period of 20 seconds were persistently recorded for most earthquakes, so this was chosen as a reference. This is equivalent to describing the loudness of a piece of piano music by measuring only the strength of sounds with a pitch of, say, middle C. If most music had a lot of these C notes and all tones were played with the same loudness, such a system might provide a reasonable estimate of how loud the sound was. But if one were to play the C softly and hammer away on the low notes, the sound would be much louder than indicated by measuring only notes with a pitch of C. Such proved to be the case when reporting earthquake magnitude using only the strength of 20-second seismic waves—the quakes could really be much more powerful than the Richter scale indicated.

When more modern seismograph instruments called broadband seismographs, with the ability to measure a much broader spectrum of earthquake frequencies, were developed, it became apparent that some earthquakes had much of their energy at the lower, slower frequencies. Seismologists have termed such earthquakes "slow earthquakes," and many of these last for a very long time compared to other quakes. The speed with which the earth actually ruptures during a typical subduction zone earthquake might be from 5,500 to 7,500 miles per hour (2.5–3.5 km/sec). A slow earthquake might rupture at an abnormally slow 2,000 miles per hour (1 km/sec) and last from 60 to 130 seconds, two to four times longer than a typical earthquake. Such a slow release of seismic energy generates few of the higher frequency seismic waves with a period of 20 seconds. As a result, much of the energy of slow earthquakes was "missed" when the magnitude was measured using the Richter scale. It was also discovered that as quakes got larger, they sent out a greater proportion of their energy in the longer period waves and that the waves measured for surface magnitude never indicated magnitudes greater than 8.0 to 8.3, no matter how large the earthquake. In other words, the Richter scale becomes saturated during large earthquakes, actually beginning to saturate at around 7.3. This meant that the Richter scale was also underestimating the true magnitude of very large earthquakes as well as slow earthquakes.

Fortunately, seismologists had developed other ways of describing the energy released by an earthquake. One of the most reliable and consistent was a measure known as "seismic moment." Seismic moment describes the faulting process itself and considers such factors as the area of the fault, the length of the displacement, and the strength of the rock being ruptured. Seismic moment was then used to produce a new magnitude scale M_W, referred to as "moment magnitude."

How would the moment magnitude of tsunami earthquakes compare with their Richter magnitude? Could this explain the generation of large tsunamis from small earthquakes? The Nicaragua earthquake in 1992 was the first tsunami earthquake to be measured by a network of modern broadband seismic instruments. The earthquake had been assigned a Richter magnitude of 7.0, but when the moment magnitude was calculated, it turned out to be a significantly larger 7.6. It was also a slow earthquake, taking nearly 2 minutes for the seismic energy to be released as the Cocos tectonic plate thrust under the North American plate. It has since been estimated that the Sanriku earthquake of 1896 might have had a Richter magnitude of only around 6.8, but was probably a slow earthquake, with a rupture estimated to have taken as long as 100 seconds, a period over which it released enough energy to create giant tsunami waves.

The Ansei-Nankai Earthquake Tsunami

It now appears likely that an earthquake that struck Japan on December 24, 1854, was another slow earthquake. Shortly after the event, a correspondent of the *New York Herald* who was in Shanghai recorded the following account from the first officer of the Russian frigate *Diana*, which had been in the port of Shimoda when the earthquake and tsunami struck. "At 9:30 A.M. the sea was observed washing into the bay in one immense wave, thirty feet high, with awful velocity; in an instant the town of Shimoda was overwhelmed, and swept from its foundations." The *Diana* was a total loss.

Another account comes to us in a letter to Commodore Perry, who had recently visited Shimoda for the ratification of the treaty between the United States and Japan. Captain Adams of the U.S. Navy, who had witnessed the events at Shimoda, wrote: "The sea rose in a wave five fathoms [30 feet] above its usual height, overflowing the town and carrying houses and temples before it in its retreat. . . . Only 16 houses were left standing in the whole place. The entire coast of Japan seems to have suffered by this calamity."

Yet another account comes from the Japanese village of Hiro in Wakayama. Translated from a Japanese primary school text by Dr. O. Muta of Murdoch University, Australia, it tells the story of the residents and the village squire named Gohei.[1] The story, known as the "Fire of Rice Sheaves," is one of the most poignant accounts to come out of any tsunami.

> "It is not normal," Gohei muttered to himself as he came out of his house. The earthquake was not particularly violent. But the long and slow tremor and the rumbling of the earth were not of the kind old Gohei had ever experienced. It was ominous.
>
> Worriedly he looked down from his garden at the village below. Villagers were so absorbed in the preparation for a harvest festival that they seemed not to notice the earthquake.
>
> Turning his eyes now to the sea, Gohei was transfixed at the sight. Waves were moving back to the sea against the wind. At the next moment the expanse of the sand and black base of rocks came into view.
>
> "My God! It must be a tsunami," Gohei thought. If he didn't do something, the lives of four hundred villagers would be swallowed along with the village. He could not lose even a minute.
>
> "That's it!" he cried and ran into the house. Gohei immediately ran out of the house with a big pine torch. There were piles of rice sheaves lying there ready for collection. "It is a shame I have to burn them, but with this I can save the lives of the villagers." Gohei suddenly lighted one of the rice sheaves. A flame rose instantly fanned by the wind. He ran frantically among the sheaves to light them.
>
> Having lit all the sheaves in his rice field, Gohei threw the torch away. As if dazed he stood there and looked at the sea. The sun was already down and it was getting dark. The fire of the rice sheaves rose high in the sky. Someone saw the fire and began to ring the bell of the mountain temple.
>
> "Fire! It is the squire's house!" Young men of the village shouted and ran hurriedly to the hill. Old people, women and children followed the young men. To Gohei, who was looking down the hill, their pace seemed as slow as ants. He felt impatient. Finally about twenty young men ran up to him. They were going to extinguish the fire. "Leave them! There

[1] M. I. El-Sabh, The role of public education and awareness in tsunami hazard management, in *Tsunami: Prediction, Disaster Prevention and Warning,* ed. Y. Tuschiya and N. Shuto, pp. 277–285.

> will be a disaster. Have the villagers come here." Gohei shouted in a loud voice. The villagers gathered one by one. He counted the old and young men and women as they came. The people looked at the burning sheaves and Gohei in turn.
>
> At that time he shouted with all his might. "Look over there! It is coming." They looked through the dim light of dusk to where Gohei pointed. At the edge of the sea in the distance they saw a thin dark line. As they watched, it became wider and thicker, rapidly surging forward.
>
> "It is a tsunami!" Someone cried. No sooner than they saw the water in front of them as high as a cliff, crashing against the land, they felt the weight as if a mountain was crushing them. They heard a roaring noise as if a hundred thunders roared all at once. The people involuntarily jumped back. They could not see for a while anything but clouds of spray which had advanced to the hill like clouds.
>
> They saw the white fearful sea passing violently over their village. The water moved to and fro over the village two or three times. On the hill there was no voice for a while. The villagers were gazing down in blank dismay at the place where their village had been. It was now gone without a trace, excavated by the waves.
>
> The fire of the rice sheaves began to rise again fanned by the wind. It illuminated the darkened surroundings. The villagers recovered their senses for the first time and realized that they had been saved by this fire. In silence they knelt down before Gohei.

The "long, slow tremor" felt by Gohei seems to describe a slow earthquake with its energy at the lower frequencies. This earthquake would have probably had a small Richter magnitude but a large moment magnitude. Earthquake magnitude based on seismic moment is now considered to be a much more reliable indicator of the ability of an earthquake to generate a tsunami than Richter magnitude. In fact, one eminent seismologist and tsunami expert, Dr. Emil Okal of Northwestern University, has gone so far as to state, "the exclusive use of magnitudes (Ms) for tsunami warnings is dangerous, if not outright criminal."

Other Tsunami Amplifiers

Many of the so-called tsunami earthquakes have now been satisfactorily explained by the difference between their Richter and moment magnitudes. However, there are still a few tsunamis that remain too big even when the earthquake magnitude is calculated using their seismic moment, and other explanations have been presented by scientists. It has been suggested that earthquakes striking ocean trenches might generate tsunamis up to ten times larger if part of the earthquake energy is propagated into thick ocean sediment deposits. Sedimentary layers at the ocean floor are much more elastic than the typical rocks that make up the ocean crust. As a result the wedge-shaped sedimentary masses found in some trenches might increase motion of the sea floor by as much as three times that of an ordinary thrust earthquake and hence produce much larger tsunamis.

Another mechanism has been proposed to explain an anomalously large tsunami generated during the Torishima earthquake of June 13, 1984. In this case a volcanic mechanism has been proposed. It is suggested that a mixture of volcanic magma and water was suddenly injected into the sediments, increasing the movement of the sea floor and thereby generating a larger tsunami.

It has also been suggested that some tsunamis may have been large compared to their generating earthquakes because the earthquakes weren't solely responsible for the tsunami. This explanation invokes submarine landslides triggered by the earthquake to account for at least some of tsunami generation. Such a mechanism has been proposed for the 1929 Grand Banks earthquake and tsunami (described in Chapter 2) and the 1946 Aleutian event (Chapter 1). There is no direct proof of a large submarine landslide off Unimak Island in 1946. However, a tsunami in Fiji in 1953 occurred following a relatively small earthquake and there is substantial proof that the waves were produced by a submarine landslide triggered by the quake.

The 1953 Suva, Fiji, Earthquake and Tsunami

Fiji has been struck by 11 tsunamis, or "*loka*" in Fijian, during the past century. Thanks to the shallow seas surrounding the islands and their protective barrier reefs, only one of these has caused significant damage and resulted in loss of life. It occurred at 12:26 P.M. on September 14, 1953, when a Richter magnitude 6.7 earthquake struck Fiji with an epicenter only about 10 miles southwest of the capital of Suva. Many small landslides occurred on the hillsides of the main island of Viti Levu and small waves came ashore on Beqa and Kadavu islands almost immediately following the earthquake. Meanwhile in the city of Suva as many as a hundred people watched as sea level began to drop at the entrance to the harbor. Next a big "brown bubble" surfaced, followed by an enormous surge "boiling up in the harbor entrance." Thousands of people in Suva looked on as the tsunami approached, clearly visible from as far as two miles away. The wave, estimated to have been as high as 50 feet, broke as it struck the outer edge of the barrier reef, tossing enormous chunks of coral up onto the reef flats.

All along the south coast the tsunami struck with great violence, throwing the hulk of a sunken vessel up on to the reef five miles south of Suva, and at Beqa pitching up blocks of coral 10 feet in diameter. The villages of Nukulau, Makuluva, and Nukui, not protected by barrier reefs, suffered serious damage. The village of Nakasaleka on Kadavu Island is also not protected by a barrier reef and is situated at the head of a small bay. When the first small waves came ashore immediately after the earthquake, most of the villagers took them as a warning and fled to higher ground, but two village elders refused to leave. Several heroic villagers stayed behind trying to convince the old men to evacuate. Only 30 seconds had passed when a large tsunami wave, estimated at 15 feet high, struck the village. The two elders were drowned, but the others managed to survive by climbing nearby trees.

Back in Suva, the wave spread out once inside the harbor and was only 6 feet high by the time it crashed into the waterfront and broke against the seawall. This was high enough, however, to claim three more lives, bringing the death toll to five. Fortunately, it was low tide. Had it been high tide, the tsunami might well have turned into a major catastrophe.

Several pieces of evidence point to a submarine landslide as the source of the tsunami. In addition to the small size of the earthquake, the tsunami struck both Suva and Kadavu within seconds after the earthquake, far too quickly for a tsunami generated above the epicenter of the earthquake to have traveled to both locations. The most likely origin of the tsunami was an area along the steepest portion of the coastal shelf extending from Suva to Beqa. Here the bottom abruptly drops off into deep water. The initial drop in sea level at Suva was a clear indication of slumping as the edge of the reef broke off and slid down slope. A bathymetric survey carried out following the earthquake revealed that the depth just beyond the edge of the broken coral reef had increased by at least 100 feet.

Evidence of the extent of underwater landslide activity was provided by the breakage of two out of three submarine cables from Suva. The cable to Norfolk Island, passing through the Kadavu Passage directly beneath the slide area, was the most seriously damaged. Over 60 miles of cable was either buried or carried away, with the piece south of Beqa being found some 13,000 feet away from its original position. An old steel hawser, blasted clean of rust by the sediment-charged current, was found tangled in the cable. The evidence strongly suggested that massive landslides along the south coast not only generated tsunami waves but also gave rise to a powerful turbidity current. The total volume of material thought to have slumped following the earthquake was estimated to be as great as 350 million cubic yards. This was probably not the first such enormous submarine landslide to strike Fiji. A recent bathymetric survey has revealed huge slumps and submarine canyons along the entire southern coast of Viti Levu. Like the Hawaiian Islands, the islands of Fiji appear to have been sculpted by massive submarine slides and slumps, which may have generated enormous tsunamis.

Though non-seismic tsunami generation mechanisms have been less well studied than seismic mechanisms, tsunamis generated by landslides and volcanoes appear to occur repeatedly in the same locations. The monitoring of active volcanoes has received increased attention since the explosion of Mount St. Helens, while the tsunami threat they pose has been largely ignored. But it is the tsunamis from landslides for which we are the least prepared, and our safety could benefit from research and mitigation measures. Landslide-generated tsunamis provide no advance warning such as we receive from an earthquake or volcanic eruption. However, by studying the areas where tsunami-generating landslides have occurred and noting areas with similar slopes and sediment conditions, it should be possible to predict those areas in the greatest danger. Coastal areas where glaciers enter the sea are often at risk. Most fjords have steep sides, which can collapse during a small earthquake. Glacial deltas and outwash plains are usually mixtures of poorly sorted sediments, which tend to be unstable. Glaciers themselves may have vertical or overhanging fronts, which can surge and collapse into bays. These conditions pose a definite risk for local tsunamis. But not only glaciers create the conditions for submarine landslides. As previously mentioned, major submarine landslides have struck areas ranging from sparsely settled Newfoundland, to the tourist resorts of the French Riviera and the Hawaiian Islands. It now appears that another prime tourist destination could be at risk from tsunamis. Puerto Rico and the Virgin Islands are located on a highly seismic plate boundary which, though not well understood, appears to have little vertical motion. This would suggest little risk from earthquake-generated tsunamis. However, sea floor images from GLORIA have identified a cliff along the northern slope of Puerto Rico that is most likely to have been produced by a large submarine landslide. The steep slopes and high seismicity that produced the landslide are still present in the region, hence there is the potential for the generation of a giant catastrophic tsunami in this part of the Caribbean.

Other Tsunami Phenomena: Sound, Light, and Edge Waves

Sound

In addition to the various types of seismic waves that travel through the earth, earthquakes also produce sound waves. The acoustic energy from submarine earthquakes is known as the "T-phase" and represents the sound created by the motion of the sea floor. Where would we collect such data? Who would be interested in listening for sounds in the ocean?

The U.S. military built large arrays of underwater microphones called "hydrophones" during the Cold War to determine the splash-down points for missiles being tested by other nations. From September 1982 through April 1990, T-phase was recorded by three large hydrophone arrays on the Pacific Ocean floor near Wake Island. Put to a peacetime use by Drs. Chip McCreery and Dan Walker of the University of Hawai'i, the recordings were analyzed for T-phase. The scientists determined that within a certain frequency range (10–35 Hz), the strength of the T-phase corresponded to the seismic moment of the earthquake. The loudest T-phase (that is, the most energy) was produced by earthquakes that generated measurable tsunamis, whereas earthquakes that did not generate tsunamis produced quieter T-phase. In other words, earthquakes generating tsunamis were very noisy underwater, almost as if the sea floor was crying out with high-pitched screaming whenever tsunami waves were created.

This would seem a very promising avenue of research for determining when tsunamis are generated. Unfortunately, the T-phase is also strongly influenced by the shape of the sea floor as the sound waves travel from their source to an array of hydrophones. Islands, bits of continents, and underwater mountains and ridges all strongly influence the T-phase, making its interpretation extremely complex and site specific.

Light

An even more unusual way of studying tsunamis was suggested by Dr. Walker in Hawai'i following the 1994 Kuril Island tsunami. At Punalu'u on the northeast shore of the island of O'ahu, Delores Martinez had heard the tsunami warning and evacuated her coastal home for higher ground. As she stood with her neighbors at a lookout point about 165 feet above sea level waiting for the tsunami, she began to make a video recording of the ocean. On her recording a dark "shadow" can be seen at the horizon extending across the entire field of view. After a couple of minutes, the shadow has moved down from the horizon and the bright ocean surface appears behind the shadow. As the shadow begins to approach shore, in the background on the tape can be heard, "Look at that son-of-a-gun coming. . . . Seven hundred miles an hour." The shadow was obviously visible to the naked eye, and the observer was clearly impressed by the speed with which it was approaching the shore. Next a radio can be heard in the background with the announcer saying, "It's already 10:44 and the wave was supposed to hit at 10:42"—the shadow was right on schedule.

As the shadow continued to approach shore, the observers became so concerned that they turned off the camera and climbed to higher ground. From their perspective it was impossible to judge the height of any waves, but the size and speed of the shadow were terrifying. From the video the shadow appears to be about a mile wide and to be traveling at a speed of about 100 miles per hour. The shadow continued to approach the coast until about half a mile offshore it encountered the shallow reef and disappeared. Over the next 90 minutes, three additional "shadows" were observed at about 20-minute intervals. All swept in from the horizon with great speed, but none was as wide as the first one.

The same phenomenon was noted elsewhere on O'ahu by observers at elevation above the shore. Two Honolulu Civil Defense officers, David Kinolau and Craig Huish, were standing on a bluff overlooking the ocean on the north shore of O'ahu when they noticed a shadow moving toward them and then disappear near shore. Susan Kennedy was in her Honolulu apartment about 260 feet above the sea level. She was looking southwest toward the shore at the time the tsunami was due to arrive when she noticed the shadow near the horizon, with the leading and trailing edges appearing as straight lines parallel to the horizon. Ian Walters was in his home south of Diamond Head overlooking the ocean from a height of about 200 feet when he noticed a "shadow appearing out of nowhere." It also extended completely across the horizon in a straight line, and as it approached shore, it disappeared upon striking the reef.

What was this mysterious shadow? Dr. Walker believes that it was produced as the tsunami wave "deformed the ocean's surface." This could affect the amount of light reflected or radiated toward an observer. It is not known whether the shadow was produced by the tsunami crest, trough, or face of the advancing wave, but Dr. C. D. Mobley, an expert on optical effects and water, suggests: "This is the same thing you see when a gust of wind makes a 'cat's paw' on the water surface. On a clear day, the cat's paw appears darker than the undisturbed water because it is reflecting a deeper-blue sky light. . . . I think the effect on the video is purely one of sky light being reflected differently by the sea surface as the tsunami passes by."[2]

Dr. Walker suggests that the phenomenon might be "detected and measured with optical devices in low-flying aircraft or satellites" and used for tsunami warning purposes. Indeed, it may have been this same phenomenon that was observed by the pilot of the PBY flying north of O'ahu on the morning of April 1, 1946.

Light effects of tsunamis have also been observed at night, when they are produced by an entirely different mechanism. On March 3, 1933, the

---

[2]Walker, D. A. 1996. Observations of Tsunami "Shadows": Technique for Assessing Tsunami Wave Heights? *Science of Tsunami Hazards* 14:3–12.

Sanriku coast of Japan was struck in the middle of the night by a disastrous tsunami with waves as high as 75 feet. A remarkable aspect of this tsunami was that some of the waves were "luminous" as they approached shore. A strong light seemed to be emitted from the surface of the water as tsunami waves struck near the mouth of Kamaishi Bay. The most logical explanation for this phenomenon is that the turbulence created by the advancing wave front stimulated tiny marine organisms called dinoflagellates, which have the ability to create biological light, or "bioluminescence." The dinoflagellates simultaneously "turned on" in response to the tsunami, producing the strong flash of light witnessed by observers.

Edge Waves

On April 25, 1992, a Richter magnitude 7.1 earthquake occurred off the coast of northern California near Cape Mendocino. Uplift occurred on the sea floor, and a tsunami was generated. Small tsunami waves were recorded along the coasts of California and Oregon and as far away as Hawai'i.

At Trinidad, about 60 miles north of the epicenter, two fishermen had close encounters with the tsunami. One was trying to launch his boat from his car trailer, when "a six-foot wave ran up under his car." The second was trying to load his boat on his trailer when it was swamped by the wave. Both fishermen had to have their cars and trailers towed out of the sand.

What was especially interesting about this tsunami was that at several locations, two separate sets of tsunami waves came ashore. Crescent City, Arena Cove, North Spit, and Point Reyes all measured a first "packet" of tsunami waves arriving as would be predicted by their travel times through deep water; a second packet arrived much later. At Crescent City the first packet arrived 47 minutes after the earthquake and the second came 155 minutes after the earthquake. The same situation occurred at other sites along the coast. At North Spit, near Eureka, the first wave arrived in 20 minutes and the second group in 2½ hours; at Arena Cove the first packet came in at 35 minutes and the second at 3 hours and 30 minutes; and at Point Reyes the first tsunami struck about 65 minutes after the quake and the second some 3 hours after the shock. Dr. Frank Gonzalez, at the Pacific Marine Environment Laboratory in Seattle, has carried out an in-depth study of this tsunami. According to Dr. Gonzalez, the second set of tsunami waves "represent trapped coastal waves" called "edge waves." Edge waves move parallel to the shore and have their maximum size at the coastline. Since they are moving in shallow water, they travel much more slowly than deep water tsunami waves. But this doesn't necessarily mean that they don't carry plenty of energy. The first tsunami waves that struck Crescent City were only about 14 inches high, whereas the edge wave packet that arrived later surged up over 21 inches, nearly twice as high. Reports by fishermen suggest that the tsunami may have surged even higher, to about 4 feet, in the inner harbor at Crescent City.

These facts are not just of academic interest; they have important implications for tsunami warnings. An area threatened by a tsunami may be first struck by small waves; then, feeling the danger has passed, communities may be off guard and unexpectedly inundated by larger waves hours later. Not only do coastal communities need to worry about rapid evacuation, they must also be prepared for an extended evacuation period.

There was another very disturbing aspect to the Cape Mendocino earthquake and tsunami. The tsunami was generated by an earthquake occurring just off the U.S. west coast. This was not a tsunami coming from Alaska, Japan, or far-off Chile. Could a really large earthquake occur here and generate giant destructive tsunami waves? What was the tsunami threat from close at hand?

Tsunami Threat to the U.S. West Coast

Most of the great tsunamis generated along the west coast of the Americas are produced as the Pacific tectonic plate is thrust beneath the North and South American plates. Off the U.S. west coast, however, the dynamics of the plates are quite different and rather complex. From near Puerto Vallarta, Mexico, north along the coast of California to Cape Mendocino, the two plates do not collide but instead slide past each other. This movement is responsible for the well-known San Andreas Fault and a number of lesser faults. These faults are mostly on land; consequently the majority of the earthquakes produced by the motion (93 percent) have their epicenters on land. The dominant movement is horizontal (strike-slip), which usually does not generate tsunamis. However, a few of the faults have produced vertical movement offshore, which could generate tsunamis. Once such case occurred on November 4, 1927, when a Richter magnitude 7.3 earthquake struck the edge of the continental shelf. A tsunami was generated that was recorded as far away as the Hawai-

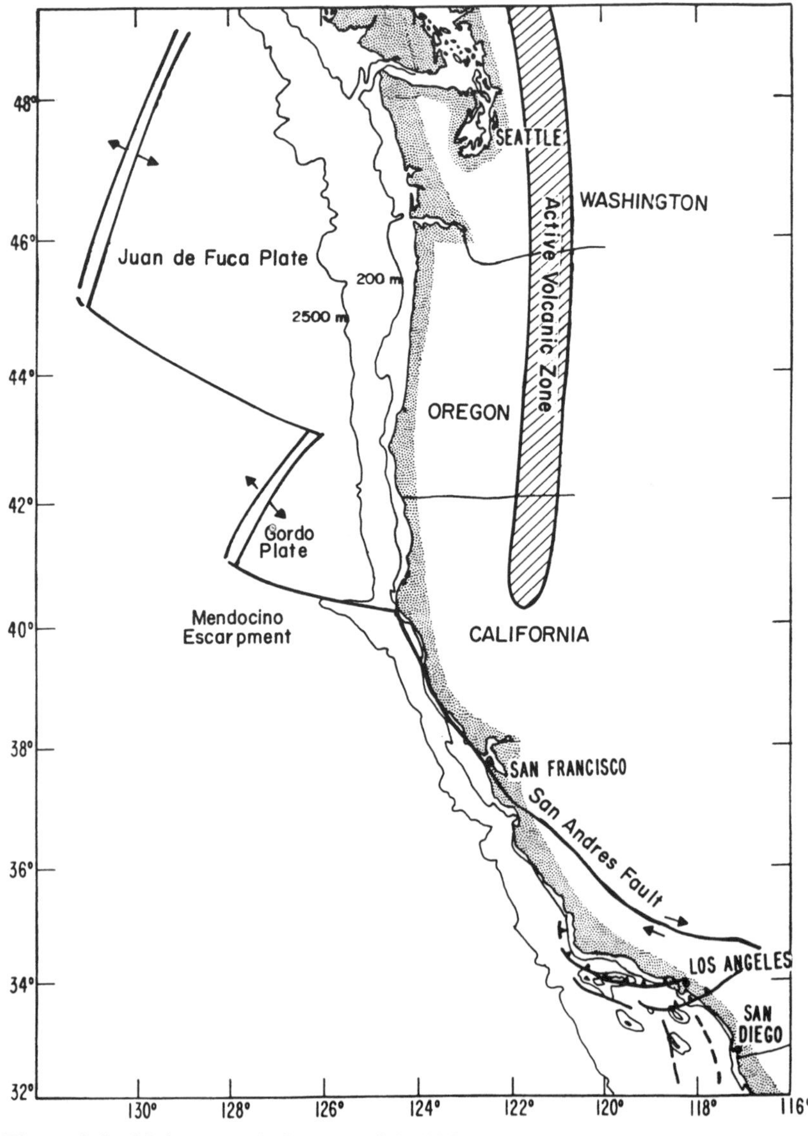

Figure 9.4 Major tectonic features of the U.S. west coast.

ian Islands and had significant wave heights along the coast of southern California. At the town of Surf a worker on the Southern Pacific Railway reported that a wave about 6 feet high came ashore, and at Port San Luis, another railway employee reported a 5-foot rise and fall of the sea. The tsunami was also observed at Pismo Beach, where it was compared to large storm waves.

More recently, on October 17, 1989, a Richter magnitude 7.1 earthquake ruptured a section of the San Andreas Fault centered in the Santa Cruz Mountains of central California. The earthquake resulted in vertical uplift of the sea floor as well as submarine landslides along the continental shelf off Monterey Bay and along the walls of the Monterey Submarine Canyon. The combination of submarine slides and vertical motion produced a small tsunami that struck the coast of central California. At Santa Cruz harbor, less than 10 miles from the epicenter, the harbormaster observed water begin rushing out of the harbor almost immediately after the earthquake and continuing for 15 to 20 minutes. At the marine laboratory at Moss Landing, water began to drain from the Old Salinas River about 10 minutes after the tremor. About 20 minutes after the earthquake the tsunami reached Monterey Bay. The water in the bay then began to move up and down with a range of about 16 inches every 9 minutes as the tsunami triggered a seiche. Neither of these tsunamis caused much damage or resulted in loss of life, and there is little historical evidence of extensive tsunami inundation from locally generated tsunamis in the region, though this is no guarantee against future tsunami impacts.

Farther north, however, the situation is very different. From Cape Mendocino north to the northern limit of Puget Sound, the arrangement of the tectonic plates becomes even more complicated. Offshore lie the small Gorda and Juan de Fuca tectonic plates. These small oceanic plates are converging with the North American plate at the rate of about an inch a year along what is known as the Cascadia Subduction Zone. Most subduction zones are marked by the presence of deep oceanic trenches and the occurrence of very large earthquakes—the kinds of areas that gave us the devastating 1946 and 1960 tsunamis. The Cascadia Subduction Zone, however, lacks a deep ocean trench and has produced no giant earthquakes in historical times. Why should this subduction zone be so different? Some geologists have postulated a different type of subduction, one without earthquakes, or alternatively a subduction zone that is simply dying. But there is little hard evidence for either hypothesis. The third and most frightening option is that the subduction zone is stuck or "locked," with the stress still accumulating until it is released in a really large earthquake. The locked subduction zone has been dubbed the "megathrust" and has frightening implications. Some geologists have speculated that it might require a "sequence of great earthquakes (magnitude 8) or a giant earthquake (magnitude 9)" to release the pent-up stress and unlock the megathrust.[3]

The 1992 Cape Mendocino earthquake was especially important because it was the first large historical earthquake to be measured on the Cascadia Subduction Zone and hence provided the first direct evidence that the subduction zone was active. The details of the quake provide further support for active subduction. The rupture began onshore at a depth of about 6 miles and propagated toward the surface and seaward. The main shock was followed the very next day by two considerable aftershocks, both located offshore and measuring 6.6 on the Richter scale. The earthquake sequence was recorded on 14 different seismographs within a radius of 80 miles, and these registered some of the strongest ground shaking ever recorded. The quakes resulted in $60 million in damage, destroying 202 buildings and damaging another 906. At least 356 people were injured. The toll would have been much greater had the quake not been centered in a sparsely populated area where most people live in small, wood frame houses, which tend to stand up rather well to earthquakes. The main shock produced uplift, ranging from 20 inches to nearly 5 feet, along 15 miles of coast. Other observations and calculations based on complex seismic parameters indicate that there was nearly 9 feet of "thrust" along the megathrust.

Many scientists are now convinced that this was a subduction zone thrust earthquake, occurring at the extreme southern end of the 700-mile-long subduction zone. But what of the huge seismic gap to the north? Is there any evidence of a large earthquake elsewhere along the subduction zone? Tsunami expert Dr. James Lander has recently discovered historic documents indicating that a previous large quake may have occurred near

---

[3]L. Dengler and K. Moley, *Living on shaky ground* (California Office of Emergency Services, Federal Emergency Management Agency, and Humboldt State University, 1995)

the California-Oregon border in 1873. The following accounts appeared in *The Crescent City Courier* on November 29, 1873, and spoke of an earthquake and tsunami across the state line in Oregon near Port Orford.

> A loud noise was heard off at sea West of Cape Blanco. It appeared like the rush and upheaval of the waters; in fact the water was seen to rise and fall, boiling and hissing. This took place or was noticed immediately after the shock and the people in that vicinity were making preparations for climbing a tree or getting to higher ground. No tidal waves followed and nothing unusual noticed on the beach.
>
> N.B. Mr. Deadmond who resides one mile north of here directly on the sea coast says that about ten minutes after the shake he heard a noise off to the westward, loud as the report of a hundred cannons, and that he noticed indications on the beach of very high water and sand being thrown up to the highest water marks.

Another account appeared in *The Humboldt Times* of Eureka, California, on December 1, 1873 and relates an account from Crescent City.

> [A]n earthquake shock exceeding in violence anything ever before experienced here since the settlement of the country. . . . The shock was strong enough to ring the town firebell and to set all the door bells in the town ringing. All the brick buildings in town and crockery-ware suffered considerable damage. The Indians here were very frightened and took to the hills and highlands. It seems they have a tradition that such an occurrence took place here once before, followed by what, as they describe it, must have been an immense tidal wave.

These accounts point to a significant earthquake occurring in the region of the Cascadia Subduction Zone, quite possibly generating a tsunami. Even more intriguing are the comments about an Indian tradition of an "immense tidal wave." There are, in fact, several Pacific Northwest Native American legends that may describe tsunamis. The following account comes to us from a 1929 interview with Agnes Mattz, a full-blooded Tolowa woman, forty-five years old.

> This happened in Oregon in a place called Brookings now, but Chetko is the Indian name. It is a very pretty place on the bank of the river. There was an old woman who was so old that she was blind. She had two grandchildren, a boy and a girl.
>
> All of a sudden, the grandmother told the children to go right away,

> to go as fast as they could and not to wait for anybody. She would stay. She was old and ready to die anyway.
>
> The two children ran as fast as they could, upstream, away from the harbor toward Mount Emilie as their grandmother had told them. Halfway there they looked back. They could hear people cry. They could hear the cries rise and sink out. They could see the water come.
>
> When the sun came up it [the water] had all gone away. Everything was gone. They went back to where their house had been. There wasn't anything there, no dead people. Everything was swept clean. The ocean was nice and smooth. The boy went down to the beach to fish. He saw far away someone slowly coming toward him. It was a girl. He went to meet her. They got married and people started again.

Another story, told to Judge James Swan by a chief and published in 1868, is as follows:

> A long time ago, but not at a very remote period, the water of the Pacific flowed through what is now the swamp and prairie. . . . The water suddenly receded leaving Heeah Bay perfectly dry. . . . It . . . then rose again . . , til it had submerged . . . the whole country, excepting the tops of the mountains . . . as [the water] came up to the houses, those who had canoes put their effects into them, and floated off with the current. . . . and when the waters assumed their accustomed level, . . . many canoes came down in the trees and were destroyed, and numerous lives were lost.

Though there are, indeed, some inconsistencies in these legends, what they describe is not inconsistent with a large tsunami. But wouldn't such an event leave some physical evidence?

Dr. Brian Atwater of the U.S. Geological Survey believes that he has found just such evidence. In the Seattle area around Puget Sound, there are signs of a large earthquake and tsunami around 1,000 years ago. Radiocarbon dates made on submerged trees in landslide deposits in Lake Washington indicate simultaneous slides at three sites at this time. Another site about 6 miles northwest of downtown Seattle and one on Whidbey Island, 20 miles farther south, contain sand deposits with plant material also dated at around 1,000 years old. Similar deposits at Grays Harbor, Willapa Bay, and at the Copalis River estuary also indicate subsidence from an earthquake followed by inundation by tsunami waves between 1,000 and 1,100 years ago.

The same deposits also contain evidence of a more recent event about

Figure 9.5 Deposit left by ancient tsunami at Willapa Bay, Washington. The dark layer below the center of the photograph is the layer of sand transported and deposited by the tsunami waves.

300 years ago. At this time the forest apparently subsided into the intertidal zone and "a tsunami left sand on the freshly subsided lowlands." The sand deposits become thinner away from the ocean and contain fossil remains of tiny marine plants called diatoms—all strong evidence supporting a tsunami as the source of the deposit. What produced the subsidence and tsunamis? In the words of Dr. Atwater, "The events can be explained most simply as great thrust earthquakes on the boundary between the Juan de Fuca and North American plates."

But wouldn't such large earthquakes be expected to produce Pacific-wide tsunamis? Though there are few reliable written records in the Pacific dating back 1,000 years, an event 300 years ago should have been recorded. In fact, there are Spanish records of tsunamis from an earthquake in Peru in 1687 and another in Chile in 1730, but no record of a tsunami from a distant source around the year 1700. In Japan, however, Dr. Kenji Satake and others have found evidence of a tsunami striking the Pacific coast of Japan around midnight on January 27, 1700. Tsunami waves between 6 and 10 feet high struck the village of Miyako, damaging 20 houses. About 20 miles farther south, the village of Otsuchi was hit by waves up to 10 feet high, and at Tanabe, 600 miles southwest, the tsunami struck with waves estimated at about 6 feet high. There is no record of an earthquake in Japan on this date. The relatively uniform height of the waves, between 6 and 10 feet at all three villages, and the lack of an earthquake suggest a tsunami from a distant source as opposed to one generated near Japan. To further test this scenario, Dr. Satake has produced a computer model showing that a magnitude 9 earthquake off the Pacific Northwest would produce a tsunami with run-up heights of about 6 feet on the Pacific coast of Japan. This is further supported by Native American legends of a large earthquake striking the Pacific Northwest on a winter's night. If the tsunami that struck Japan at midnight on January 27 were generated by an earthquake on the Cascadia Subduction zone, it would have occurred around 9 P.M. on January 26—a winter's night.

In spite of the fact that a major earthquake has not been scientifically documented off the coast of Washington and Oregon, it now appears likely that they have occurred previously and that the Cascadia Subduction Zone is a prime candidate for a large thrust earthquake capable of generating a destructive tsunami in the future. In the words of Washington State Geologist G. W. Thorsen, "The written history of the region is too short to define what is geologically normal, as indicated by the recent catastrophic eruption of Mount St. Helens."

Other Seismic Gaps

There are other subduction zones around the Pacific Rim that in some ways resemble the Cascadia zone. These are found off southern Chile, southwestern Japan, and Colombia. Though there may be some scientific doubt as to future earthquakes and tsunamis in these regions, there are other areas about which there is little doubt as to their tsunami potential.

Some of the most worrisome gaps occur off Alaska. Estimates of how often earthquakes reoccur in the region vary from as frequently as every 60 years to as infrequently as every 1,000 years, but there is a general consensus, based on the motion of the tectonic plates, that large earthquakes should repeat at least every 200 years for the region as a whole. There are three gaps that have had no large earthquakes this century. The 1957 earthquake ruptured only a small segment of the Aleutian Arc and did not rupture the area off Unalaska Island. A large earthquake in 1938 off the Alaska Peninsula did not extend into the Shumagin Islands area, and an earthquake there on May 13, 1993, is considered too small to have "filled the gap." In addition to the Shumagin gap off the Alaska Peninsula, and the Unalaska gap in the central Aleutians, the Yakataga gap has been identified in the Gulf of Alaska. Large tsunamis have been generated in each of these areas in the past and are certain to occur again in the future—though just how soon is the question. A study published in 1984 indicates that a major earthquake (moment magnitude of 8.4) could occur in the Shumagin gap sometime within "the next two decades." Another cause for concern is the directional nature of a tsunami generated in the Shumagin gap. Many scientists believe that a Shumagin Islands tsunami "would be directed toward the south and southeast." To paraphrase University of Hawai'i geophysicist Gerard Freyer: instead of getting waves that spread out in all directions, like dropping a pebble in a pond, for the Aleutians it's more like rolling a log into the pond, with the waves headed straight to Hawai'i. Tsunamis from this area consequently have great potential for producing damaging waves in Hawai'i and along the west coast of the United States and Canada.

Other Tsunami Threats: Bolide Tsunamis

A tsunami of giant proportions, with wave amplitudes as great as 100 meters (330 feet), is referred to as a "mega-tsunami." Mega-tsunamis are thought to have been occasionally produced by massive submarine landslides, such as that which occurred about 100,000 years ago in the Hawaiian Islands. But another mechanism could generate a tsunami of this size or even larger.

There is evidence on land of as many as 100 large extraterrestrial objects striking the earth during the past 300 million years. Though many of the impact structures, called "astroblemes," are questionable, at least 13 of them were definitely produced by large meteorites, asteroids, or small comets, collectively referred to as "bolides." Examples include the Barringer crater in Arizona and several well-preserved craters in Australia. Since the earth is 70 percent ocean, for every bolide that strikes land, there should be roughly three impacts in the ocean, or "hydroblemes." The water itself will not sustain a crater, and any crater formed on the sea floor would soon be covered by marine sediments and ultimately disappear beneath an ocean trench as the oceanic tectonics plates are subducted. But there is another type of evidence. A bolide striking the ocean would create a mega-tsunami of immense proportions. Though the actual size of the tsunami depends on the diameter of the bolide, the depth of water, and a number of other factors such as atmospheric pressure, the resulting tsunami could be as high as the water is deep. For a deep ocean site this could mean tsunami waves up to 3 miles high near the point of impact. As the water cavity created by the impact collapsed, a series of waves could be produced. At a distance of 3,000 miles from the impact site, waves could still be as high as 300 feet. Such giant waves would be expected to wash up huge deposits of material along the margins of the continents. Recently just such evidence has been found off the Atlantic coast. Drilling into the ocean floor off Virginia uncovered a 200-foot thick layer of boulders buried beneath 1,200 feet of other sediments. Among the boulders were found glassy rock fragments called tektites and mineral grains with so-called shock features—both evidence of a bolide impact. It is theorized that a large meteorite stuck the continental shelf here about 40 million years ago, producing the tektites and shocked minerals. A tsunami then surged across the shallow sea floor, piling up boulders 3 feet in diameter into a layer 200 feet thick.

An even more impressive impact site lies farther south. About 65 million years ago the last of the dinosaurs and thousands of smaller and less well-known creatures suddenly became extinct. Most earth scientists now believe that this massive extinction event was caused by an asteroid up to 6 miles in diameter, striking somewhere in the Caribbean. The most likely site is an area of the Yucatan Peninsula of Mexico called the Chicxulub impact site. The impact of an asteroid of this size would have the energy equivalent to a 100-megaton bomb and could be expected to have produced a prodigious tsunami. A site in northeastern Mexico contains a bed of tektites nearly a foot thick, along with a layer of boulders, coarse-grained sandstone, wood, and the skeletons of deep-water marine organisms. This deposit is thought to have been created by a 150- to 300-foot tsunami that washed deep water materials onto the land, with the drawback of the wave carrying terrestrial vegetation into the ocean.

The impact of such a large bolide could have truly global implications. Blast and earthquake damage would occur on a regional scale, producing fires and dust that could alter the earth's climate. A larger impact, from, say, a moderate-sized comet, would not only create a gigantic tsunami but also send a splash of water from the initial impact to a height as great as 25 miles. The injection of sea water at this altitude could have serious consequences for the atmosphere, leading to complex chemical reactions, with the chlorine from sea water possibly causing destruction of the earth's protective ozone layer.

Dr. Charles Mader, an expert in creating computer models of waves, has produced a model to determine the tsunami inundation of the Hawaiian Islands by an asteroid impact in the Pacific Ocean. He used the recent impact of the Shoemaker-Levy-9 asteroid with Jupiter as a typical event, and made calculations for an asteroid 3 miles in diameter traveling at 43,000 miles per hour. Such an impact in the Pacific Ocean could be expected to generate tsunami waves that would inundate the Hawaiian Islands to heights of between 500 and 1,000 feet above sea level.

Fortunately, the impact of such large bolides has been relatively rare in Earth's history, most impacts being from much smaller objects, but even small bolides are thought to be capable of causing enormous damage primarily by their tsunami waves. Calculations indicate that an asteroid only 1,300 feet in diameter could create tsunami waves capable of flooding up to 3,000 feet of the coastal plain around an entire ocean basin.

Though the ocean impact of a bolide has never been observed, calculation can be made using mathematical techniques, some of which were originally developed to study the impact of nuclear explosions. At the height of the Cold War there was actually some concern about the possibility of tsunamis generated by the detonation of nuclear bombs off the continental shelf of the U.S. east coast. Fortunately, a nuclear bomb has never been detonated on the continental shelf, but a huge explosion on the east coast did generate a tsunami. The blast occurred in midst of World War I and was the largest explosion in history up until that time. It took place in the narrows of the Canadian port of Halifax, Nova Scotia, on December 6, 1917, when just before 9 A.M. the munition ship *Mont Blanc* collided with the relief ship *Imo*. A fire broke out on board the *Mont Blanc* and the crew wisely and quickly abandoned ship. The *Mont Blanc* was carrying a full cargo of explosives equivalent to 2,900 tons of TNT. The ship began drifting across the harbor toward the city of Halifax, where it grounded near one of the piers and then exploded with tremendous force. An entire section of the city was demolished and a small steam ship was blown out of the water and over the top of one of the piers. The casualty toll would reach 2,000 dead and 9,000 injured. The damage produced by the explosion would later be studied during the Manhattan Project as a means of estimating the blast produced by the first atomic bombs.

The explosion also generated a destructive tsunami, which may have been responsible for many of the casualties. Near the site of the explosion, survivors reported seeing three "tidal waves"; one was "more than twenty feet above the level of the harbor" and "emptied Halifax Harbour right out. . . . just a little stream of water down the middle on the bottom." In the harbor but away from the narrows, the tsunami run-up was probably in the range of 6 to 12 feet. Here Lucy Ann Slaunwhite's father sat in a boat moored at the downtown waterfront. According to her account of his experience, "the tide lifted the boat right up over the wharf to which they were tied."

Within the narrows, the water reached much higher levels, achieving its maximum run-up at Dartmouth, just opposite the explosion. Phillip Mitchell was on the Dartmouth shore at the time of the explosion. He ran to an electric pole, and seeing "a huge wave," "he wrapped his arms about the post and held on." The water surged over his head and soaked a boxcar on the railroad track above him, but he managed to hold on and survive. The wave then receded, only to be followed by a second smaller wave, which fortunately "only reached his arm pits."

Tufts Cove, home of several families of Micmac Indians, also felt the full impact of the tsunami. Several houses were destroyed and nine people were killed by the waves. George Dixon and his brother-in-law were working in a shipyard on the cove when they were knocked off their feet by the explosion. When they got to their feet and saw the tsunami wave approaching, George's brother-in-law commented, "We might as well have been killed as to find this thing [the tsunami] coming at us."

Improvements in Tsunami Warnings

The ability to provide a timely warning of a gigantic tsunami produced by an asteroid crashing into the ocean may be many years away, but major advances are now being made in improving the warning system for tsunamis generated by earthquakes. As we have seen, determining the tsunami-generating potential of an earthquake by using moment magnitude is more accurate than using surface magnitude. It would seem obvious to adopt moment magnitude in order to provide more accurate warnings and reduce false alarms. But the determination of moment magnitude not only requires data from broadband seismograph instruments, which have not been available at most seismic stations, but is a rather difficult procedure. To quote Northwestern University seismologist Emil Okal, "the moment is the amplitude of the deviatoric stress tensor, a mathematical object of dimension 5, whose resolution necessitates the inversion of a large amount of seismic data." As complex as this sounds, seismologists are capable of quickly carrying out the necessary calculations, but it can take up to an hour to collect enough data to solve for all the components of the "moment tensor." In order to be able to provide a rapid estimate of seismic moment for the purpose of predicting the tsunami potential of earthquakes, geophysicists working at the French tsunami warning center, the CPPT (Centre Polynesian de Prevention des Tsunamis) in Tahiti, developed a technique for estimating the moment magnitude using the data collected by a broadband seismograph at a single station. This rapid estimate of seismic moment omits the exact depth and geometry of the earthquake focus, as obtaining this information requires waiting for data from other stations, but appears to provide a reliable real-time measure of an earthquake's tsunami-generating potential. Their technique is known as TREMORS for *T*sunami *R*isk and *E*valuation through Seismic *Mo*ment from *R*eal-time *S*ystem. Testing TREMORS in 1986, the CPPT correctly determined that the May 7 earthquake was not big enough to generate a tsunami. The Pacific Tsunami Warning Center, however, using primarily surface wave magnitude, issued a warning for Hawai'i and the U.S. west coast that resulted in a costly and embarrassing false alarm. In 1992, the CPPT, which used TREMORS to evaluate the Nicaraguan earthquake, was the only warning center that correctly identified its tsunami potential. Had the TREMORS system been in operational use by the warning system, an alert could have been provided to coastal areas of Costa Rica and Nicaragua at least 22 minutes before the tsunami struck. If the system had been used during the Java earthquake and tsunami, a warning could have been issued at least 18 minutes before the tsunami struck there. The TREMORS system has since been put into use by the regional warning center in Chile, and finally in 1997 it was adopted by the Pacific Tsunami Warning Center in Hawai'i.

The regional warning system in Chile also tested another advance in providing rapid tsunami warnings through a project called THRUST (*T*sunami *H*azards *R*eduction *U*tilizing *S*ystems *T*echnology). Project THRUST created a reliable, low-cost local tsunami warning system that makes use of satellite communication technology but also uses existing tsunami warning methods. The advances of the project have now been fully integrated in the National Tsunami Warning System of Chile, and it is hoped it will provide a faster and better response to future tsunami events there.

The Japan Meteorological Agency has continued to improve its tsunami warning system. It now telemeters seismic data to each of six Regional Tsunami Warning Centers in Japan where the data are quickly analyzed and tsunami forecasts are issued 7 to 8 minutes after an earthquake. Warning information is disseminated by the regional centers using dedicated telephone lines to avoid the communication problems experienced in 1993. Following the 1993 tsunami, a new seismograph network containing 150 broadband seismographs was constructed and the JMA now makes use of the moment magnitude in identifying which earthquakes are most likely to generate tsunamis.

The West Coast/Alaska Tsunami Warning Center has also made major improvements in recent years in expanding its monitoring capabilities and in the area of integrating computers into its operations. It continuously monitors seismic data from 34 short period seismometers across Alaska and the lower 48 states and from 30 broadband instruments, all of which are routed directly to its computers for real-time processing. It also monitors via satellite 8 real-time National Ocean Survey tide sites, as well as over 80 near real-time tide sites throughout the Pacific, for immediate confirmation of water levels during a tsunami watch.

During several recent earthquakes with tsunami generation potential, the center managed to issue warnings in an average of 11 minutes from the time of the quake, a reduction of more than 50 percent from the time previously required to provide a warning. Its communications system has also been improved with use of the National Weather Wire Satellite (NWWS) communication system. The NWWS is now used to disseminate information on tsunami watches and warnings to Alaska, Canada, Washington, Oregon, and California.

The combined capabilities of the Tsunami Warning System (TWS) of the Pacific are now quite impressive. A large number of countries and territories that had not been members of the system in 1960 joined soon after the destructive Chilean tsunami. There are now tide stations between South America and Hawai'i, including Kanton Island (Kiribati), Johnston Island (United States), Baltra Island (Ecuador), and Easter Island and San Felix Island (Chile). Satellite telemetered tide stations have been installed at some 25 sites across the Pacific. These stations operate on their own independent power sources, secure from electrical outages resulting from earthquakes. Sea level measurements are made every 2 seconds, averaged over a 3 or 4 minutes interval, and the data routinely transmitted by satellite to the Pacific Tsunami Warning Center (PTWC) every 3 to 4 hours. In the event of a tsunami wave, however, an "event detector" almost instantaneously sends a message to the PTWC over a special emergency satellite channel. Though satellite coverage is limited at present to the southwest and south Pacific, including South America, future satellite data collection platforms will be used to help cover the entire Pacific region. The new satellite-telemetered tide stations go a long way in helping to fill the gap in confirming the generation of tsunamis from the coast of South America.

As with the regional warning centers, the operational center for the

TWS, the Pacific Tsunami Warning Center, has seen its capabilities improved and operations expanded. In 1996 a new computer system was installed, providing a badly needed upgrade to the information-handling capacity of the center. As of 1998 the PTWC had real-time access to seismic data from nearly 900 seismic stations and monitored approximately 100 tide stations. They could disseminate tsunami watch and warning information to some 50 different Pacific basin nations and island entities, ranging from the Russian Federation and the People's Republic of China to Brunei and Pitcairn Island.

Measuring Tsunamis in the Deep Sea

The use of broadband seismographs and the TREMORS system goes a long way toward improving the accuracy of predicting the tsunami potential of an earthquake, but this system cannot confirm whether or not a tsunami has actually been generated by an earthquake. Since the inception of the warning system, confirmation has been provided by tide stations. But tide stations are not without their problems. First of all, tide stations can provide data "only after a tsunami has arrived." This means that tide station sites are areas that must be "sacrificed" to collect data—not exactly desirable from the viewpoint of the tide station. Second, the harbors in which gauges are situated complicate the tsunami signal they record, because the period of seiche of a harbor as well as its overall shape may increase or decrease the amplitude of the tsunami. Furthermore, until recently the gauges used for measuring the tides were specifically designed not to measure waves. This became obvious when observers repeatedly reported tsunami wave heights more than twice those recorded on nearby tide gauges. Though tsunami waves have a slower frequency than wind-generated ocean waves, their period is much faster than that of the tides; consequently much of the tsunami record may be filtered out or "clipped" by the tide gauge. The filtering effect is often produced simply by the small size of the hole allowing water in or out of the stilling well of the gauge. The shorter the period of the tsunami wave, the worse this filtering effect becomes. This problem has now been largely eliminated by a new generation of tide gauges, known as the New Generation Water Level System (NGWLS). These gauges use acoustic techniques to measure the water level in the stilling well and record tsunami waves more faithfully. In addition, new computer software has been designed to quickly detect the presence of a tsunami and automatically send an emergency message to the warning center via satellite transmission. This software was installed on the tide station at Adak, Alaska, just a week prior to a small tsunami on June 1, 1996, and it "performed flawlessly."

However, there is still a problem with even these newest gauges. The range of measurement possible with any tide gauge is limited to the length of the stilling well. If the amplitude of a tsunami wave is larger than the maximum tidal range for which the gauge was designed, the water level may be higher and/or lower than the top and bottom of the stilling well. This will result in a "clipping" of the tsunami wave record. All of these considerations may be irrelevant, though, as tide gauges are almost always mounted on piers with their associated electronic and communication gear housed in a nearby shed. Any reasonably destructive tsunami will destroy the gauge and shed, not to mention compromising any real-time emergency communication capability.

In any event, the data from a tide station near the area where a tsunami is generated may be of questionable value. Tsunami waves are often strongly amplified near their source, and hence the record provided by a tide station near the origin of a tsunami may not give an accurate measure of the size of the waves being propagated across the ocean. Another problem is the directionality of the tsunami waves. Normally the highest waves will be directed perpendicular to the long axis of the area of uplift or depression, usually following the axis of a subduction zone. An earthquake in the Aleutian Islands, for example, will direct most of its energy toward the mid-Pacific. Much less energy will be directed toward the sides. Hence the tide stations located on other Aleutian Islands may record smaller waves, which could be misinterpreted to indicate a small tsunami headed toward the mid-Pacific.

The best tide station data for providing a record of tsunami amplitude not complicated by local bathymetric effects are those from isolated oceanic islands such as Midway and Wake Island. There are many such islands in the south and western Pacific, but there are fewer in the eastern Pacific—and no such islands between the Aleutians and Hawai'i, or the Aleutians and the west coast of the United States and Canada. In these areas tsunamis warnings must be based on confirmation from tide stations close to the source, stations that may be providing questionable data. This is the very procedure that has resulted in so many false alarms.

In fact, false alarms are the greatest single threat to the credibility of the tsunami warning system. Since 1948, there have been 20 tsunami warning in Hawai'i resulting in evacuation of coastal areas. Of these, 15 have been considered false alarms. No emergency system can maintain public credibility with a 75 percent false alarm rate. Add to this the lack of reliable island tide stations to the north of Hawai'i, and you have a system that is practically doomed to failure for tsunamis generated in much of the north Pacific. But there may be a solution on the horizon. As early as 1975, geophysicist Martin Vitousek had the idea of putting tsunami monitoring gauges on the ocean floor in the north Pacific. At that time the technology to create such a system was either not available or not affordable, but in recent years the necessary components for such a system have become available practically "off the shelf." The heart of the system is a bottom pressure recorder (BPR), a device that sits on the ocean floor and can measure the tiny changes in water pressure produced as tsunami waves travel overhead. In 1986 testing was begun on an array of four BPRs deployed in the north Pacific. Though there were no Pacific-wide tsunamis during the test period, one of the devices was located only about 600 miles away from the epicenter of a 1988 earthquake. About 70 minutes after the earthquake, it successfully recorded small tsunami waves traveling at about 500 miles per hour. These BPRs, deployed and recovered by an oceanographic vessel, had to be retrieved in order to download their data. The Japanese have tested their own BPRs deployed offshore of Japan; they transmit the data via submarine cable back to the JMA laboratories. Their system has successfully recorded small tsunamis from earthquakes in the Mariana Trench in 1990 and again in 1993. Japanese scientists remarked that the tsunami records resembled those from tide stations on small isolated islands, that is, they lack the modifying effects of local bathymetry.

Then in 1995, a prototype operation system was tested off the coast of Washington and Oregon. This setup differed from the previous systems in that a surface buoy was moored above the BPR. Data from the BPR are sent through the water from the sea floor as sound waves. The acoustic signal is picked up at the surface buoy by a special receiver (an acoustic modem). The data are then transmitted to a satellite high overhead that relays the signal to the warning center. During the test deployment, the

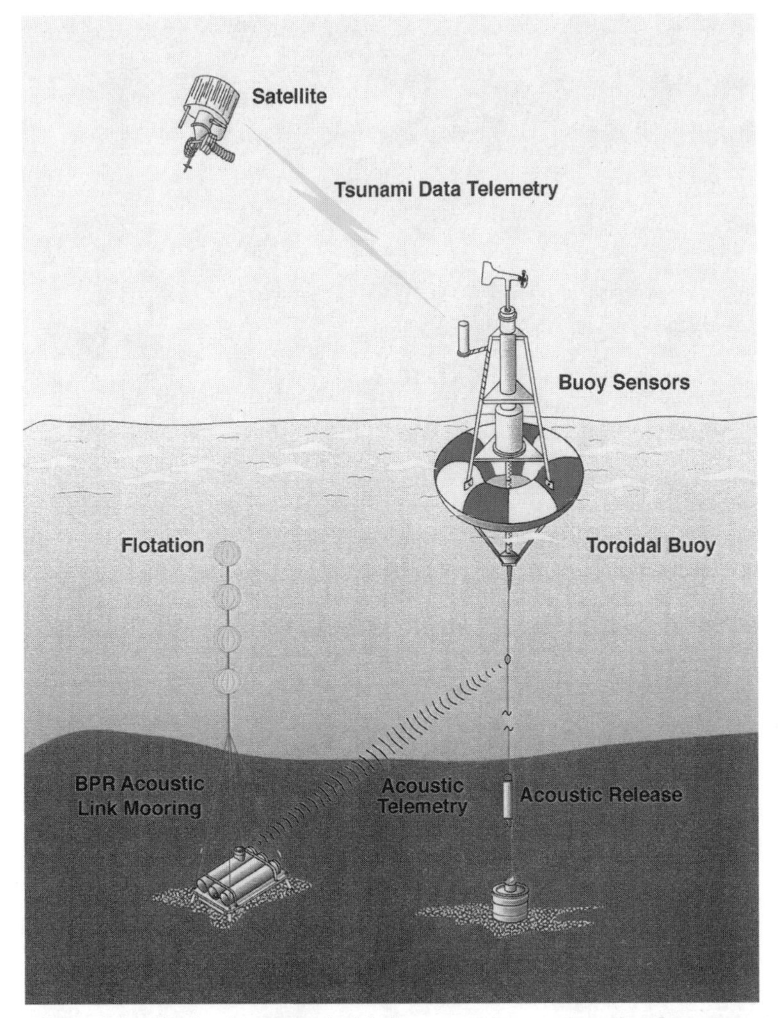

Figure 9.6 Design for tsunami real-time reporting system using deep-water tsunami detectors.

surface buoy was subjected to waves over 20 feet high, yet the system was able to record and transmit data even under these rough sea conditions. A tougher test is to see if the system can survive the stormy waters of the U.S. Congress.

In July of 1997 the first "deep ocean tsunami detection and real-time reporting system" was deployed in the Gulf of Alaska south of the Shumagin Islands, and in September 1997 a second system was put into

Figure 9.7 Deployment of the surface buoy for a deep-water tsunami detector off the Oregon coast from the NOAA ship *Ronald H. Brown*.

position off the coast of Washington and Oregon. These BPRs sit on the sea floor in water over 15,000 feet deep. Four more systems are scheduled to be deployed by the year 2001, at a cost of about $2 million. Five systems, including the ones deployed in July and September, are to be spread out across the north Pacific from Kamchatka to Oregon. A sixth system is to be deployed in the eastern equatorial Pacific ocean. This last buoy will provide Hawai'i with vital data about tsunamis originating along the coast of South America. According to Dr. Eddie Bernard, director of NOAA's Pacific Marine Environmental Laboratory in Seattle, which is in charge of the project, "though sensors don't eliminate the risk of false alarms, they certainly curb it." But, deploying, operating, and maintaining the new systems will depend on federal support. Support for tsunami warning and mitigation has always been sporadic and minimal, despite the savings that such systems could make, not only of lives, but of dollars. A single false alarm can cost from $30 to $40 million in lost salaries and revenue in the state of Hawai'i alone, not to mention the loss of credibility for the warning system. Loss of credibility can mean a subsequent increase in noncompliance during future warnings, and therefore in increased loss of life. There are also injuries and even fatalities that can occur during an evacuation due to such things as accidents and heart attacks. A few million would seem a small price to pay to reduce the number of false alarms, but unfortunately the U.S. government often seems to fail to understand the true importance of the tsunami threat. Even modest funds for tsunami research and education can be very difficult to come by. On April 1, 1996, a joint U.S.-Japan tsunami conference was held in Hilo. This date marked the fiftieth anniversary of the 1946 tsunami and also commemorated the one-hundredth anniversary of the 1896 Sanriku tsunami. The Japanese national science agency managed to provide funding to send 11 scientists to participate in the conference. The U.S. National Science Foundation, on the other hand, could not come up with even modest financial support to help sponsor the meeting.

The Shoreline Problem

Even with full deployment of deep ocean tsunami detection and real-time reporting systems, we will not have a perfect warning capability. We

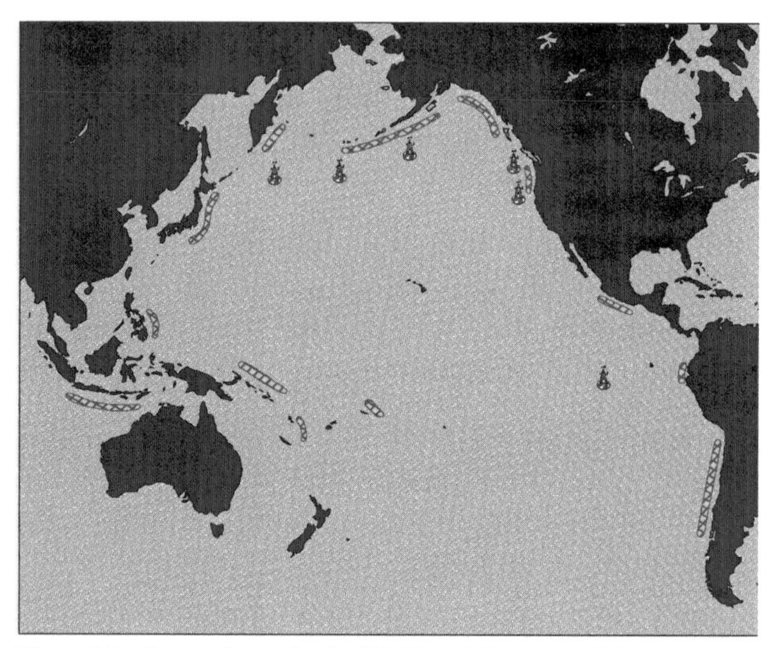

Figure 9.8 Proposed sites for the NOAA real-time tsunami detectors.

still need to learn more about how tsunamis behave when they strike a coast. Some scientists consider the problem of predicting how tsunami waves interact with the shoreline to be more difficult to solve than that of the generation and propagation of tsunamis across the ocean. Two factors must be considered when trying to estimate how a tsunami will react at the shore. First the size of the tsunami on the high seas must be known. Once deep ocean detection systems are in place, we will have largely solved this problem. The second problem is how the tsunami will respond to the shore. This problem includes any seiche effects produced in bays, in harbors, or on the continental shelf, plus the run-up effects as the tsunami waves actually climb up the shoreline. The combined impact of run-up and seiche effects can increase the size of the tsunami waves anywhere from 10 to as much as 100 times. A tsunami measuring 4 inches on the high seas could become a modest 3-foot wave in one area and a 30-foot monster in another, depending on how it interacts with the shore.

Various modeling techniques have been used to try to simulate the behavior of tsunami waves in shallow water. One such technique is called "hydraulic modeling" and uses a physical scale model for experiments.

Because of Hilo Bay's peculiar sensitivity to tsunami waves, a hydraulic model of the bay was constructed following the 1960 tsunami. The model measured 85 by 62 feet and represented the triangular shape of Hilo Bay. Model tsunami waves were produced by releasing water from large tanks according to a program.

Experiments with this hydraulic model have shown that almost any wave that enters the bay either directly hits downtown Hilo or is bounced off the coast to the north of the town. Sometimes the direct and reflected waves interact constructively to produce especially large waves in the center of the bay near Coconut Island.

With today's high-speed computers, it is possible to calculate these effects by producing a mathematical simulation of the tsunami waves in a

Figure 9.9 Hydraulic model of Hilo Bay. The horizontal scale is 1:600 and the vertical scale is 1:200. The lines shown on the model indicate the limits of inundation of the tsunamis of 1960 (dashed line), 1957 (dotted line), and 1946 (dashed and dotted line). The commercial docks can be seen at the top of the picture. The section of the breakwater just to the left of the docks was carried away during the 1946 tsunami. The railroad bridge across the Wailuku River can be seen in the lower right-hand corner of the picture. The bridge was destroyed during the 1946 tsunami.

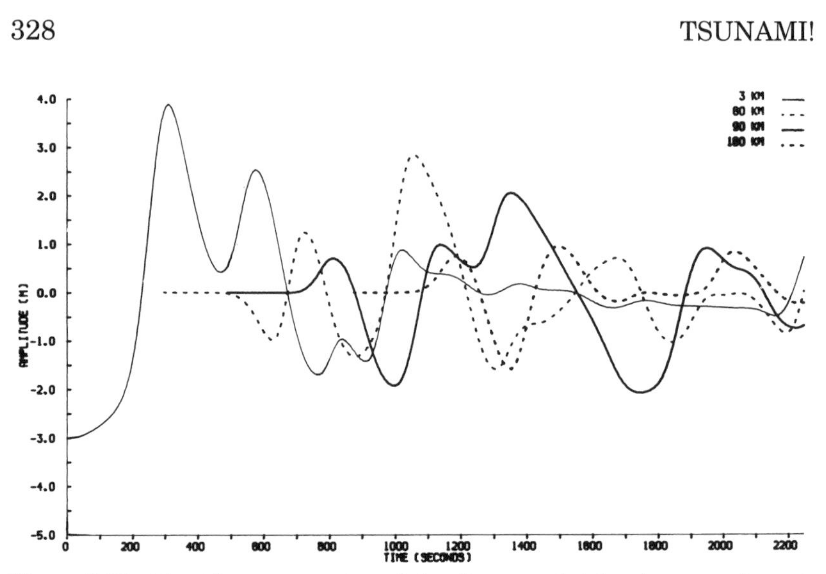

Figure 9.10 Graphic representation of shoreline wave heights from a mathematical model of the 1975 Hawaiian tsunami.

"numerical model." The models, however, are only as good as the data upon which they are based. These data can come only from measurements and observations of real tsunamis in coastal waters.

Coastal Tsunami Monitoring

The best way to learn about tsunamis is, of course, to study the tsunami waves themselves. Because tsunamis are fortunately not everyday events, it is imperative to collect as much information as possible from each occurrence. In Hawai'i this task has been coordinated under the Tsunami Monitoring Program directed by the University of Hawai'i.

When a tsunami watch is declared, trained volunteer observers, professional scientists, and cooperating military personnel begin preparing their equipment. By the time a tsunami warning goes into effect, they are ready to put the monitoring program into operation. Observers head toward preselected shoreline vantage points. Here, 8-millimeter time-lapse surveillance cameras, as well as video and still cameras, are set up to film the waves. At sites safe from the advancing tsunami, observers stay to operate the equipment and note the wave activity. At other, less secure, sites the cameras are set to run automatically after the area is evacuated.

Self-contained portable tsunami gauges can be deployed from piers and in designated shoreline areas. These gauges sense the change in water pressure as the waves pass over them and record their measurements on a computer chip. U.S. Navy P-3B patrol planes take to the air prior to the arrival of the first waves. The aircraft fly at an altitude of 1,000 feet in a racetrack pattern over critical shoreline areas, covering the same spot every 15 minutes. Special cameras mounted in the belly of the plane take 480 exposures on 70-millimeter film, documenting the arrival of each tsunami wave on the shores of the Hawaiian Islands.

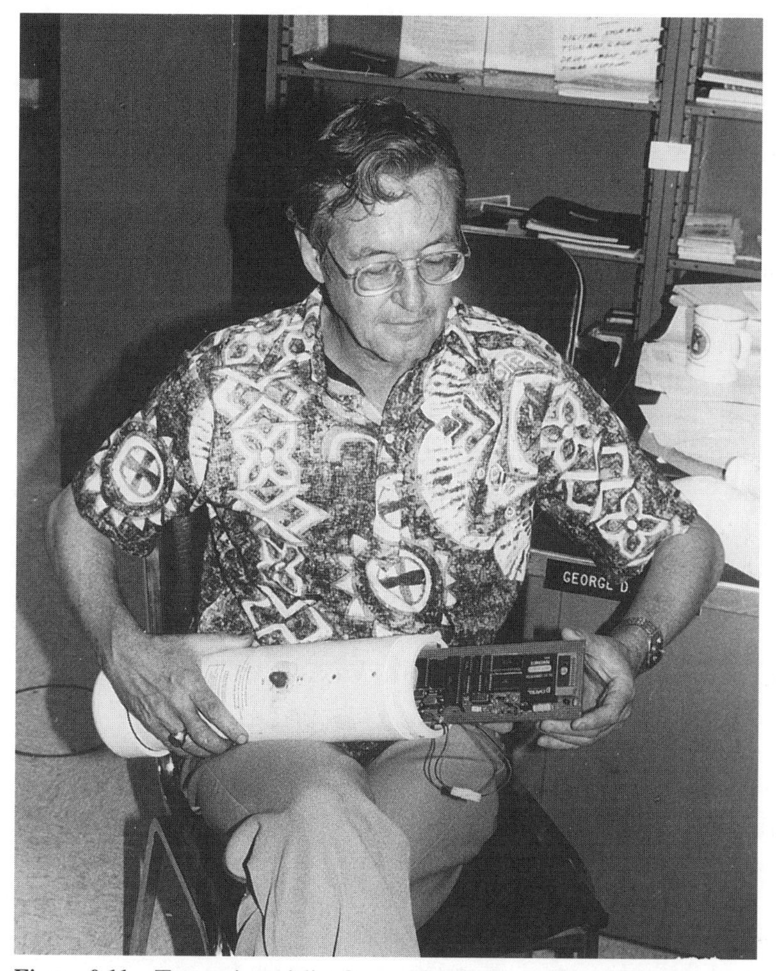

Figure 9.11 Tsunami specialist George Curtis displays a prototype portable gauge for measuring tsunami waves.

After the tsunami waves have passed and it is once again safe to enter shoreline areas, the portable tsunami gauges and surveillance cameras are recovered. A ground survey and damage assessment is conducted by a team from the U.S. Army Corps of Engineers, joined by volunteers from the American Society of Civil Engineers. A post-tsunami aerial photographic survey is undertaken jointly by the U.S. Navy and Coast Guard with the Civil Air Patrol, and even makes use of special National Weather Service "U-2" surveillance aircraft. To remain in a state of readiness, the Tsunami Monitoring Program holds yearly meetings and conducts rehearsals of most aspects of the program.

There are even independent efforts to collect data on tsunami run-up in the Hawaiian Islands. Tsunami expert Dr. Dan Walker has personally produced and installed around the Hawaiian Islands what must be one of the most cost-effective "tsunami run-up gauges" ever devised. His water level measuring system consists of a plastic gutter pipe filled with empty beverage cans. The gauge is mounted to a vertical pole or tree in an inundation zone and sits and waits for the next tsunami. When a tsunami wave does arrive and flood the area, the cans fill up as the water rises inside the pipe. In order to measure the run-up level after the tsunami, Dr. Walker merely begins drilling holes through the pipe into the cans, starting at the top and working his way down the pipe, until he strikes water.

Monitoring and other types of tsunami research will help us to understand better the threat posed by tsunamis, but there are also important measures that can be taken to mitigate the destruction from the next tsunami.

Mitigation Measures

In Hilo following the devastation of the 1946 tsunami, the idea of protective walls and breakwaters was considered as a possible defense against tsunamis. Cost estimates indicated that to be effective such a wall would cost more than the value of all the buildings in Hilo, and the idea was quickly abandoned. But tsunami walls are used extensively in Japan. Following the 1896 Meiji tsunami, a protective wall was planned for the village of Tarou, but it took the galvanizing effect of the 1933 Showa era tsunami to actually get construction under way. The wall was completed

Figure 9.12 Simple tsunami measuring device designed and deployed at Ho'okena by Dr. Dan Walker consists of a gutter pipe filled with empty beverage cans strapped to a palm tree.

following World War II and now stands 33 feet high and 4,500 feet long. It is known as "The Great Wall" and tourists come from all over Japan to gaze on the massive structure. The cost was enormous and no one is quite sure what will happen if a tsunami wave higher than 33 feet strikes the wall.

Less expensive measures, such as planting thick rows of low trees, have been tried with some success. Trees appear to be effective in absorbing at least some of the energy of an advancing tsunami wave, and they do make a rather pleasant green border for a seaside community.

Figure 9.13 The "Great Wall" at Tarou, Japan, is designed to protect the town from tsunamis up to 33 feet high.

When all else fails, relocation may be the only option. Following tsunami devastation on the Kamchatka Peninsula of Russia, many parts of the city of Petropavlovsk were moved to higher ground to avoid future destruction. In Hilo, after being almost totally destroyed twice in 14 years, much of the downtown business district was rebuilt farther inland. The inundation zone is now largely soccer fields and public park land. Many an admiring tourist has commented on the city fathers' fine town planning to have left so much precious coastal land as green open space for recreational use. Little do they realize the real reason and true cost behind the open space.

Tsunami Education

Though some destruction may be inevitable, the loss of human life to tsunamis is not. Studies of the deadly tsunamis of the last decade have shown again and again that people died because they didn't understand the threat: they didn't understand tsunamis. Surely the most important mitigating measure needed is public education about tsunamis.

It is too late to learn about tsunamis after a warning has been declared. A recent study of the reaction to the October 4, 1994, tsunami alert on the U.S. west coast indicated that "confusion and uncertainty appeared to dominate the situation in the critical decision-making hours after a warning is issued." A survey of 14 different west coast communities found that 30 percent of the respondents felt that the "tsunami warning information was unclear or unusable," and 71 percent felt that the "information was updated too slowly." To add to the confusion, communities frequently received conflicting information and even misinformation. Though such misinformation is rarely malicious, stemming largely from ignorance, it can be deadly. For example, an article about Hilo in a recent airline flight magazine in Hawai'i mentioned tsunamis and stated, "Today electronic weather forecasting equipment prevents this kind of tragedy." Not only does this imply that tsunamis are related to the "weather" but also that they are no longer a threat to Hilo. During the 1994 tsunami alert in Hawai'i, geophysicist Gerard Feyer of the University of Hawai'i was being interviewed about the tsunami threat on local television. Following a reference to 3-foot tsunami inundation, the television newsperson commented, "But we see three-foot surf all the time. Why should anybody get excited about three-foot run-up?" The commentator obviously did not understand the difference between wind waves and tsunamis. They would, no doubt, get "excited" were they to witness a 3-foot wall of water coming down the street. During the 1986 tsunami, journalists in Oregon chartered helicopters to fly out to sea to "meet the wave." They refused to believe scientists who repeatedly told them that the wave would not be more than 10 inches high and even this spread over a 200-mile wavelength.

Even public officials are sometimes woefully ignorant of what the tsunami phenomenon is like. Following the 1992 Cape Mendocino tsunami, a report by the coordinator of the County Office of Emergency Planning Services stated that the "tides in Crescent City harbor fluctuated from two to four feet. . . . but no tsunamis were generated." Those 2- to 4-foot fluctuations were the tsunami. Unfortunately such misconceptions tend to be the rule rather than the exception.

There is also a certain reluctance to discuss the tsunami threat for fear of scaring away business. For example, in 1964, the tsunami expert Bernard Zetler submitted an article about the warning system to Matson Lines for publication in their company magazine. At that time the line operated passenger ships sailing between Hawai'i and California. Zetler received an apologetic note from Matson Lines saying that the article was fine, but it could not be used because the company had a "policy which prohibited publishing anything that might alarm the passengers."

This fear of discussing tsunamis is especially evident in the tourist industry and might well be dubbed the "Jaws Syndrome." In the movie *Jaws*, a giant man-eating shark devours several swimmers, yet the local community leaders refuse to allow shark warnings for fear of scaring away tourists. Something similar appears to have happened here in Hawai'i. A decade ago, the Hawai'i Visitors Bureau rejected offers of help on tsunami information for visitors. Even today, vacationers will find nothing at the airport or any car rental agency telling them what to do in the event of a tsunami warning. The visitor industry fears that the heavily publicized images of sunny beaches and swaying palms might be replaced by the specter of walls of seawater rushing through hotel lobbies. Yet ignorance could be fatal. The head of Civil Defense for the County of Hawai'i, Harry Kim, has stated that the lack of knowledge about tsunamis might well be the "kiss of death."

Education about a natural disaster does not need to be frightening. After all, the greatest fear most people have is that of an unknown threat. The fact that hurricanes periodically strike Florida has not hurt the tourist industry there as much as has the relatively remote possibility of being shot at in a rental car. Tourists and the general public like to be informed and know what risks they are facing.

Education about tsunamis is best started early while children are still receptive to accepting the tsunami danger as a normal part of life. Both Japan and Chile have landmark education programs about tsunamis in grade schools. The same could be done in Hawai'i. After all, we have very effective education modules dealing with fires in the home. Not only do these lessons teach the student about the threat and the correct response to the danger, but children bring home booklets that they use to produce home fire evacuation plans for their families. Both of the authors have been badgered by their children into sitting down with them and planning a course of action in the event of a fire. The parents learn along with and from their children.

It is important that Civil Defense and the tsunami research community explain their actions to the public. This is especially important following a false alarm, when public confidence in the system is at a low point. The warning center in Alaska arranges special visits by their staff to communities after they have been evacuated during a tsunami warning and dangerous waves have failed to materialize. The scientists explain why the warning was called and stress the need for continued vigilance in face of the tsunami threat. The best warning system in the world will not be effective if the population does not know how to respond. Living in a free democracy prohibits forced evacuations. It is essential that the population understand the threat and be willing to cooperate with authorities.

The role of public education is most critical in areas that can be affected by locally generated tsunamis. Both the scientific community and those agencies involved in emergency management agree that technology alone cannot protect those in areas close to the source of a tsunami. People in a coastal area 60 miles away from the epicenter of an earthquake generating a tsunami would have less than 8 minutes to prepare from the time they felt the first seismic wave until the first tsunami waves arrived. Is this time enough to record the event properly, analyze the danger scientifically, contact emergency management authorities, and carry out evacuation procedures? Absolutely not!

This problem is particularly acute in the United States for coastal communities in Hawai‘i, along the Pacific coast of Alaska, and in the Pacific Northwest from northern California to Washington state. Some 10 million people live on top of the Cascadia subduction zone and half a million people live, work, and travel through the potential tsunami inundation zone extending from northern California through Washington state. In northern California much of the population is concentrated in low-lying coastal areas and near the estuaries of large rivers. Since 1980 the region around Cape Mendocino and the coastal areas of Humboldt and Del Norte counties has experienced 11 earthquakes with magnitudes greater than 6, with 5 of these near magnitude 7. There is increasing evidence that a large (8.5 to 9) earthquake occurred along the Cascadia subduction zone in the past and will occur again in the future. One model of a tsunami generated by such an earthquake produced waves over 20 feet high, inundating much of the Samoa Peninsula in Humboldt County and resulting in greater flooding in Crescent City than occurred during the 1964 tsunami. Another model indicates that the maximum tsunami height could exceed 65 feet along parts of the coasts of Washington and

Oregon. Worst of all, the tsunami waves would strike the coastal communities within minutes and the danger could last for up to 8 hours or longer.

Though the tsunami risk to California is greatest in the northern part of the state, the remainder of the state is not risk-free. Large earthquakes generated by movement along the San Andreas fault system might not generate a destructive tsunami themselves, but they could set off submarine landslides that could result in destructive tsunamis.

Progress is being made in public education. A videotaped public service announcement completed in 1995 was the first ever produced in the United States dealing with the threat of local tsunamis. A booklet titled "Living on Shaky Ground" dealing with the earthquake and tsunami threat was created and distributed in northern California. The early efforts appear to be having an effect. One of the questions on a survey conducted in the northern California region in 1996 asked who was primarily responsible for emergency action immediately following an earthquake. Those who had read the booklet were more likely to accept personal responsibility for the response to an emergency, not leaving it all to some government agency.

More education tools are needed, such as published evacuation maps, kits for broadcast and print media, posters, booklets, documentary films on tsunami safety with preparedness guidelines, land use and construction guidelines, and perhaps even changes in building codes. There are some existing programs and products, but in most cases these have been produced and adopted in local regions only. Tsunami education materials need to be consistent, evaluated for their effectiveness, and widely available. The greatest obstacle to a comprehensive tsunami education program has been lack of sustained financial commitment by government. Funds became available for tsunami education in northern California following the 1992 earthquake, but were a "one time expenditure." A flood of federal and state dollars after the next tsunami will be too late.

One coastal state to take the tsunami threat seriously is Oregon. In 1995 the Oregon state legislature passed two landmark laws pertaining to tsunamis. One law prohibited the construction of essential facilities such as hospitals, police and fire stations, and schools, as well as high-occupancy buildings, in tsunami inundation zones. In 1997 a Center for the Tsunami Inundation Mapping Effort (TIME) was established at the Hatfield Marine Science Center in Oregon. The goal of TIME is to accurately draw maps of potential tsunami inundation areas and help local communities plan safe locations for essential facilities like schools. Some existing facilities, such as the elementary school at Cannon Beach, Oregon, will be relocated to a more protected site. TIME is a step in the right direction, but only time will tell if continued funding is forthcoming. TIME does not expect to begin mapping the inundation zones in Hawai'i until 1999. The second Oregon law requires education programs in coastal public schools, and that the schools conduct at least three earthquake/tsunami drills per year. These are model pieces of legislation, and it would be a great advance if they were to be adopted in all areas subject to the tsunami threat. The risk from tsunamis is simply a fact of life for those who live by the Pacific Ocean. A few years of intense mitigation effort will not "solve the problem." What is needed is a long-term, institutionalized commitment to improved warnings, emergency planning, and public education.

There is one institution whose principal goal is tsunami education—the Pacific Tsunami Museum—and its founding is itself an interesting story. On the last page of the first edition of this book was a request for additional true tsunami stories. Ms. Jeanne Johnston, whose story is told in Chapter 1, contacted me (W.D.) a few years ago with the desire to share her experience and help with tsunami education. She wondered why there wasn't a tsunami museum and suggested we begin organizing such an effort. Since that day, a group of community leaders, tsunami scientists, and concerned citizens has volunteered their time and effort to organize and found the Pacific Tsunami Museum in Hilo. The goals of the museum are to provide public education about tsunamis, to preserve local history, and to promote research and study of tsunamis. The museum has been actively collecting first-hand accounts of tsunamis from survivors. Their personal experiences need to be preserved as part of the local history of the islands, but they are also a valuable educational tool to teach current and future generations about the effects of tsunami waves. The museum also sponsors statewide school tsunami essay contests, collects photographs of tsunamis, and archives the papers of tsunami specialists who have donated their materials to the museum library. A booklet on local tsunamis in Hawai'i has been produced by the museum for distribution statewide, as well as other education materials for use in schools. As of this printing the museum has a building in downtown Hilo, one that survived both the 1946 and 1960 tsunamis, and it hopes to be in full operation soon pending the availability of funds.

The Next Tsunami

The museum has its work cut out. Hawai'i is struck by more tsunamis than any other region in the world, yet less than half the 1 million permanent residents and virtually none of the 6 million annual visitors to Hawai'i understand the tsunami hazard. Despite the death and destruction experienced in this century, the state has yet to make the commitment to protection and education already undertaken by Oregon.

Some people ask just what the odds are of another big tsunami in Hawai'i. If we look at the historical records kept since 1813, we find a total of 95 tsunamis in 175 years, or about one tsunami every two years. There has been no really large Pacific-wide tsunami to strike in these islands since 1960. Does this mean that the Hawaiian Islands are long overdue? Well, events like earthquakes and tsunamis really can't be predicted by such elementary statistics. What we can be sure of is that another giant, potentially killer tsunami will come from one direction or another, one day. It may not strike tomorrow, or next week, or even next year, but there *will* be another tsunami. We will probably have adequate warning. Will the warning be understood and heeded? That is up to you.

Figure 9.14 Sign in the lobby at Uncle Billy's Hilo Bay Hotel with "tsunami evacuation instructions."

Selected Bibliography

Abe, K. 1979. Size of great earthquakes of 1837–1974 inferred from tsunami data. *Journal of Geophysical Research* 84:1561–1568.

Adams, W. M., and N. Nakashizuka. 1985. A working vocabulary for tsunami study. *Science of Tsunami Hazards* 3:45–51.

Alexander, W. D. 1891. *A brief history of the Hawaiian people*. New York: American Book Company.

Alvarez, W., J. Smit, W. Lowrie, F. Asaro, S. V. Margolis, P. Claeys, M. Kastner, and A. R. Hildebrand. 1992. Proximal impact deposits at the Cretaceous-Tertiary boundary in the Gulf of Mexico: A restudy of DSDP Leg 77 sites 536 and 540. *Geology* 20:697.

Atwater, B. F. 1987. Evidence for great Holocene earthquakes along the outer coast of Washington state. *Science* 236:942–944.

———. 1992. Geologic evidence for earthquake during the past 2,000 years along the Copalis River, southern coastal Washington. *Journal of Geophysical Research* 97:1901–1919.

Atwater, B. F., and A. L. Moore. 1992. About tsunami 1,000 years ago in Puget Sound. *Science* 258:1614–1617.

Atwater, B. F., A. R. Nelson, J. J. Clague, G. A. Carver, D. K. Yamaguchi, P. T. Bobrowsky, J. Bourgeois, M. E. Darienzo, W. C. Grant, E. Hemphill-Haley, H. M. Kelsey, G. C. Jacoby, S. P. Nishenko, S. P. Pammer, C. D. Peterson, and M. A. Reinhart. 1995. Summary of coastal geologic evidence for past great earthquakes at the Cascadia subduction zone. *Earthquake Spectra* 11:1–10.

Atwater, B. F., and D. K. Yamaguchi. 1991. Sudden, probably coseismic submergence of Holocene trees and grass in coastal Washington state. *Geology* 19:706–709.

Bennett, C. C. 1869. *Honolulu directory and sketch of the Hawaiian or Sandwich Islands*. Honolulu: Honolulu Press.

Bernard, E. N. 1991. Assessment of Project THRUST: Past, present, future. *Natural Hazards* 4:285–292.

Bernard, E., C. Mader, G. Curtis, and K. Satake. 1994. Tsunami inundation model study of Eureka and Crescent City, California. NOAA Technical

Memorandum ERL PMEL-103, 80 pp. Washington, D.C: U.S. Department of Commerce, National Oceanic and Atmospheric Administration.

Bingham, H. 1847. *A residence of twenty-one years in the Sandwich Islands.* Canadiagua, N.Y.: Goodwin.

Birnie, I. 1996. Transportation and the 1946 tsunami. *Pacific Tsunami Museum Newsletter* 1:1, 5.

Blackford, M., and H. Kanamori. 1995. Executive summary. In *Tsunami Warning System Workshop Report.* NOAA Tech. Mem. ERL PMEL-105:43–50. U.S. Department of Commerce: National Oceanic and Atmospheric Administration.

Bonk, W. J., R. Lachman, and M. Tatsuoka. 1960. A report of human behavior during the tsunami of May 23, 1960. Unpublished report. Hawaiian Academy of Sciences, Hilo.

Bourgeois, J., T. A. Hansen, P. L. Wiberg, and E. G. Kauffman. 1988. A tsunami deposit at the Cretaceous-Tertiary boundary in Texas. *Science* 241:567–570.

Bryant, E. A., R. W. Young, and D. M. Price. 1992. Evidence of tsunami sedimentation on the southeastern coast of Australia. *Journal of Geology* 100: 753–765.

Chance, G. 1973. Perceptions of earthquake-related physical events in coastal communities. In *The Great Alaskan Earthquake of 1964, Summary and Recommendation*, pp. 117–135. Washington, D.C.: National Academy of Sciences.

Coan, T. 1882. *Life in Hawaii.* New York: Randolf.

Coffman, J. L., and C. A. von Hake, eds. 1973. *Earthquake history of the United States.* Publication 41–1 (NOAA). Washington, D.C.: U.S. Government Printing Office.

Cornforth, D. H., and J. A. Lowell. 1996. The 1994 submarine slope failure at Skagway, Alaska. Proceedings of the Seventh International Symposium on Landslides, Trondheim, Norway, June 17–21, 1996.

Cox, D. C. 1987. *Tsunami casualties and mortality in Hawaii.* 55 pp. Honolulu: University of Hawai‘i Environmental Center.

Cox, D. C., and J. Morgan. 1977. *Local tsunamis and possible local tsunamis in Hawaii.* 118 pp. Hawaii Institute of Geophysics Report 77-14. Honolulu: University of Hawai‘i.

———. 1984. *Local tsunamis in Hawai‘i—implications for warning.* Hawaii Institute of Geophysics Report 84-4. Honolulu: University of Hawai‘i.

Dames and Moore. 1980. *Design and construction standards for residential construction in tsunami-prone areas in Hawaii.* Dames and Moore: Washington, D.C.

Dana, J. D. 1868. A letter of the School Inspector General in *Hawaiian Gazette*, April 29, 1868. *American Journal of Science*, series II, 46:105–123.

Davis, Nancy Yaw. 1971. *The effects of the 1964 Alaskan earthquake, tsunami and resettlement on two Koniag Eskimo villages,* 434 pp. Ph.D. dissertation. Seattle: University of Washington.

DeBois, C. A. 1932. Tolowa notes. *American Anthropologist* 34:248–262.

Dengler, L., G. Carver, and R. McPherson. 1992. Sources of north coast seismicity. *California Geology* 45:40–47.

Dengler, L., and K. Moley. 1996. Mitigating the effects of a Cascadia earthquake: The north coast California example. In Proceedings of the Pan Pacific Hazards '96 Conference and Trade Show, Disaster Preparedness Resource Centre, University of British Columbia (CD-ROM).

Doxsee, W. W. 1948. The Grand Banks earthquake of November 18, 1929. *Publications of the Dominion Observatory* 7:323–335.

Eaton, J. P., D. H. Richter, and W. U. Ault. 1961. The tsunami of May 23, 1960, on the island of Hawai'i. *Bulletin of the Seismological Society of America* 51:135–157.

Eissler, H. K., and H. Nanamori. 1987. A single-force model for the 1975 Kalapana, Hawaii, earthquake. *Journal of Geophysical Research* 92:4827–4836.

El-Sabh, M. I. 1995. The role of public education and awareness in tsunami hazard management. In *Tsunami: Prediction, Disaster Prevention and Warning,* ed. Y. Tsuchiya and N. Shuto, pp. 277–285. Netherlands: Kluwer Academic Publishers.

Finch, R. H. 1924. On the prediction of tidal waves. *Monthly Weather Reviews* (U.S. Weather Bureau) 52.

Fraser, G. D., J. P. Eaton, and C. K. Wentworth. 1959. The tsunami of March 9, 1957, on the island of Hawaii. *Bulletin of the Seismological Society of America* 49: 79–90.

Fukao, Y. 1979. Tsunami earthquakes and subduction processes near deep-sea trenches. *Journal of Geophysical Research* 84:2303–2314.

Furumoto, A. S. 1996. Using M_w or M_t to forecast tsunami heights. *Science of Tsunami Hazards* 14:107–118.

Gault, D. E., and C. P. Sonett. 1982. Laboratory simulation of pelagic asteroidal impact: Atmospheric injection, benthic topography, and the surface wave radiation field. In *The geological implications of impacts of large asteroids and comets on the Earth,* ed. I. T. Silver and P. H. Schultz. *Geological Society of America Special Paper* 190:69.

Gonzalez, F. I., E. N. Bernard, and K. Satake. 1995. The Cape Mendocino tsunami, April 25, 1992. In *Tsunami: Prediction, disaster prevention and warning,* ed. Y. Tsuchiya and N. Shuto, pp. 151–158. Netherlands: Kluwer Academic Publishers.

Gonzalez, F. I., and H. B. Milburn. 1995. Near-source tsunami measurements for forecast and warning, In *Tsunami Warning System Workshop Report,* by

M. Blackford and H. Kanamori. NOAA Tech. Mem. ERL PMEL-105: 75–78. U.S. Department of Commerce: National Oceanic and Atmospheric Administration.

Gonzalez, F. I., K. Satake, E. F. Boss, and H. O. Mofjeld. 1995. Edge wave and non-trapped modes of the 25 April 1992 Cape Mendocino tsunami. *PAGEOPH* 144:409–425.

Gordon-Cumming, C. 1883. *Fire fountains: The kingdom of Hawaii, its volcanoes, and the history of its mission*. London: Blackwood & Sons.

Greenberg, D. A., T. S. Murty, and A. Fuffman. 1993. Modeling the tsunami from the 1917 Halifax harbour explosion. *Science of Tsunami Hazards* 11: 7–80.

Gregory, J. W. 1930. The earthquake off the Newfoundland Banks of November 18, 1929. *Geographical Journal* 77:00.123–139.

Hall, D. A. 1917a. The wreck of the U.S.S. *De Soto*. In *Proceedings of the United States Naval Institute*, ed. J. W. Greenslade, 43:1151–1160.

———. 1917b. The launching of the U.S.S. *Monongahela*. In *Proceedings of the United States Naval Institute,* ed. J. W. Greenslade, 43:1161–1172.

Heaton, T. H., and P. D. Snavely. 1985. Possible tsunami along the northwestern coast of the United States inferred from Indian traditions. *Bulletin of the Seismological Society of America* 75:1455–1460.

Hebenstreit, G. T., and T. S. Murty. 1989. Tsunami amplitudes from local earthquakes in the Pacific Northwest region of North America. Part 1: The outer coast. *Marine Geodesy* 13:101–146.

Hemphill-Haley, E. 1995. Diatom evidence for earthquake-induced subsidence and tsunami 300 years ago in southern coastal Washington. *Bulletin of the Seismological Society of America* 107:367–379.

Henry, R. F., and T. S. Murty. 1995. Tsunami amplification due to resonance in Alberni Inlet: Normal modes. In *Tsunami: Progress in prediction, disaster prevention and warning,* ed. Y. Tsuchiya and N. Shuto, pp. 117–128. Netherlands: Kluwer Academic Publishers.

Hitchcock, C. H. 1911. *Hawaii and its volcanoes*. Honolulu: The Hawaiian Gazette Co.

Houtz, R. E. 1962. The 1953 Suva earthquake and tsunami. *Bulletin of the Seismological Society of America* 52:1–12.

Houtz, R. E., and H. W. Wellman. 1962. Turbidity current at Kadavu Passage, Fiji. *Geological Magazine* XCIX:57–62.

Imamura, F., and N. Shuto. 1993. Estimate of the tsunami source of the 1992 Nicaraguan earthquake from tsunami data. *Geophysical Research Letters* 20, 14:1515–1518.

Iwasaki, T. 1986. Hydrodynamic nature of disasters by the tsunamis of the Japan Sea earthquake of May 1983. *Science of Tsunami Hazards* 4:67–81.

Jaggar, T. A. 1923. Earthquake wave in Hawaii. Monthly Bulletin of the Hawaiian Volcano Observatory 11:11.

———. 1925a. Tidal waves. *The Volcano Letter* 50:1.

———. 1925b. Prediction of tidal waves. *The Volcano Letter* 51:1.

———. 1929. Seismic sea wave March 6. *The Volcano Letter* 220:1.

———. 1930. Ocean waves from submarine earthquakes. *The Volcano Letter* 274:1–4.

———. 1931. Hawaiian damage from tidal waves. *The Volcano Letter* 321:1–3.

———. 1933. Tsunami and earthquake tidal wave of March 2, 1933. *The Volcano Letter* 397:1–2.

———. 1946. The great tidal wave of 1946. *Natural History* 55:263–268, 293.

Jacoby, G. C., P. L. Williams, and B. M. Buckley. 1992. Tree ring correlation between prehistoric landslides and abrupt tectonic events in Seattle, Washington. *Science* 258:1621–1623.

Johnson, J. M., and K. Satake. 1995. Source parameters of the 1957 Aleutian and 1938 Alaskan earthquakes from tsunami waveforms. In *Tsunami: Prediction, disaster prevention and warning,* ed. Y. Tsuchiya and N. Shuto, pp. 71–84. Netherlands: Kluwer Academic Publishers.

Kanamori, H., and M. Kikuchi. 1993. The 1992 Nicaragua earthquake: A slow tsunami earthquake associated with subducted sediments. *Nature* 361: 714–716.

Kato, Y., Z. Suzuki, K. Nakamura, A. Takagi, K. Emura, M. Ito, and H. Ishida. 1961. The Chile tsunami of 1960 observed along the Sanriku coast of Japan. *Science Report, Tōhoko University 5: Geophysics* 13:107–125.

Keys, J. G. 1963. The Tsunami of 22 May 1960 in the Samoa and Cook Islands. *Bulletin of the Seismological Society of America* 53:1211–1227.

Kienle, J., K. Zygmunt, and T. S. Murty. 1987. Tsunamis generated by eruptions from Mount St. Augustine volcano, Alaska. *Science* 236:1442–1447.

Kowalik, Z., and T. S. Murty. 1984. Computation of tsunami amplitudes resulting from a predicted major earthquake in the Shumagin seismic gap. *Geophysical Research Letters* 11:1243–1246.

Lander, J. F. 1995. Alaskan tsunamis revisited. In *Tsunami: Progress, prediction, disaster prevention and warning,* ed. Y. Tsuchiya and N. Shuto, pp. 159–172. Netherlands: Kluwer Academic Publishers.

Lander, J. F., P. A. Lockridge, and H. Meyers. 1991. Subaerial and submarine landslide generated tsunamis. In *Wind and Seismic Effects*, ed. N. J. Raufaste, NIST SP 820, Proceedings of the 23rd Joint Meeting of the U.S.-Japan Cooperative Program in Natural Resources Panel on Wind and Seismic Effects. Washington, D.C.: U.S. Department of Commerce.

Latter, J. H. 1981. Tsunamis of volcanic origin: Summary of causes, with particular reference to Krakatoa, 1883. *Bulletin of Volcanology* 44-3:467–490.

Lipman, P. W., W. R. Normark, J. G. Moore, J. B. Wilson, and C. E. Gutmacher. 1988. The giant submarine Alika debris slide, Mauna Loa, Hawaii. *Journal of Geophysical Research* 93:4279–4299.

Loomis, H. G. 1976. Tsunami wave run-up heights in Hawaii. 95 pp. Hawaii Institute of Geophysics Report 76-5. Honolulu: University of Hawai'i.

———. 1979. A primer on tsunamis written for boaters in Hawaii. 8 pp. NOAA Technical Memo ERL PMEL-16. Seattle: National Oceanic and Atmospheric Administration.

Lorca, E. 1991. Integration of the THRUST Project into the Chile Tsunami Warning System. *Natural Hazards* 4:293–300.

Ma, K-F., K. Satake, and H. Kanamori. 1991. The origin of the tsunami excited by the 1989 Loma Prieta earthquake-faulting or slumping. *Geophysical Research Letters* 18:637–640.

Macdonald, G. A., F. P. Shepard, and D. C. Cox. 1947. The tsunami of April 1, 1946, in the Hawaiian Islands. *Pacific Science* 1:21–37.

Macdonald, G. A., and C. K. Wentworth. 1954. The tsunami of November 4, 1952. *Bulletin of the Seismological Society of America* 44:463–470.

Mader, C. L. 1996. Asteroid tsunami inundation of Hawaii. *Science of Tsunami Hazards* 14:85–88.

———. 1997. Modeling the 1994 Skagway tsunami. *Science of Tsunami Hazards* 15:41–48.

Mader, C. L., and G. Curtis. 1991. Modeling Hilo, Hawaii, tsunami inundation. *Science of Tsunami Hazards* 9:85–94.

Malo, D. 1898. *Hawaiian antiquities*. Honolulu: Bishop Museum Press.

Milburn, H., A. Nakamura, and F. Gonzalez. 1996. Real-time tsunami reporting from the deep ocean. PMEL Website <http://www.pmel.noaa.gov/tsunami/milburn1996.html>.

Miller, D. J. 1960. Giant Waves in Lituya Bay Alaska. Geological Survey Professional Paper 354-C. Washington, D.C.: U.S. Government Printing Office.

Mitchell, L. R. 1991. Wai pi'i aka kai 'e'e—tracking the legacy of tidal waves in Hawaiian history. *Ka Wai Ola O OHA* (September).

Miyoshi, H. 1984. Reconsiderations on the huge tsunamis—utility of the seawall. Proceedings of the 1983 Tsunami Symposium, Hamburg, ed. E. N. Bernard, pp. 107–115.

Monastersky, R. 1991. Ancient splash in the Atlantic. *Science News* 140:286.

Moore, J. G., D. A. Clague, R. T. Holcomb, P. W. Lipman, W. R. Normark, and M. E. Torresan. 1989. Prodigious submarine landslides on the Hawaiian Ridge. *Journal of Geophysical Research* 94:17,465–17,484.

Moore, J. G., and G. W. Moore. 1984. Deposit from a giant wave on the island of Lanai, Hawaii. *Science* 226:1312–1315.

Moreira, V. S. 1988. Historical and recent tsunamis in the European area. *Science of Tsunami Hazards* 6:37–42.

Murty, T. S., and S. O. Wigen. 1976. Tsunami behaviour on the Atlantic coast of Canada and some similarities to the Peru coast. *Royal Society of New Zealand Bulletin* 15:51–60.

Nemchinov, I. V., S. P. Popov, and A. V. Teterev. 1994. Estimates of the characteristics of waves and tsunami produced by asteroids and comets falling into oceans and seas. *Solar System Research* 28:260–274.

Neumann, F. 1935. *U.S. earthquakes, 1933*. Washington, D.C.: U.S. Department of Commerce.

Nof, D., and N. Paldor. 1992. Are there oceanographic explanations for the Israelites' crossing of the Red Sea? *Bulletin of the American Meteorological Society* 73:305.

Obee, B. 1989. Tsunami! *Canadian Geographic* 109:46–53.

Ogawa, T. 1924. Notes on the volcanic and seismic phenomena in the volcanic district of Shimabara, with a report on the earthquake of December 8, 1922. *Memoirs of the College of Science,* Kyoto Imperial University, series B, 1:201–254.

Okal, E. A. 1988. Seismic parameters controlling far-field tsunami amplitudes: A review, 1988. *Natural Hazards* 1:67–96.

Okal, E. A., and J. Talandier. 1986. T-wave duration, magnitudes and seismic moment of an earthquake: Application to tsunami warning. *Journal of the Physics of the Earth* 34:19–42.

———. 1989. M_m: A variable-period mantle magnitude. *Journal of Geophysical Research* 9:4169–4193.

Oppenheimer, D. H., G. Beroza, G. Carver, L. A. Dengler, J. P. Eaton, L. Gee, F. Gonzales, M. Magee, G. Marshall, M. Murray, R. C. McPherson, B. Ramanowicz, K. Satake, R. Simpson, P. Somerville, R. Stein, and D. Valentine. 1993. The Cape Mendocino, California, earthquake of April 1992: Subduction at the triple junction. *Science* 261:433–438.

Papazachos, B. C., and P. P. Dimitriu. 1991. Tsunamis in and near Greece and their relation to the earthquake focal mechanisms. *Natural Hazards* 4: 161–170.

Pelayo, A., and D. Wiens. 1992. Tsunami earthquakes: Slow thrust-faulting events in the accretionary wedge. *Journal of Geophysical Research* 97:15,321–15,337.

Pospichal, J. J. 1996. Calcareous nannofossils and clastic sediments at the Cretaceous-Tertiary boundary, northeastern Mexico. *Geology* 24:255.

Powers, H. A. 1946. The tidal wave of April 1, 1946. *The Volcano Letter* 491:1–4.

Ramirez, J., H. Titichoca, J. F. Lander, and L. S. Whiteside. 1997. The minor destructive tsunami occurring near Antofagasta, Northern Chile, July 30, 1995. *Science of Tsunami Hazards* 15:3–22.

Reid, H. F., and S. Taber. 1920. The Virgin Islands earthquake of 1867–1868. *Bulletin of the Seismological Society of America* 10:9–30.

Rielinger, D. M. 1991. Remnants of an ancient wave. *Sea Frontiers* 37:9.

Roberts, E. B. 1961. History of a tsunami. In *Smithsonian Report for 1960,* pp. 327–340. Washington, D.C.: Smithsonian Institution.

Roberts, J. A., and C.-W. Chien. 1964. The effects of bottom topography on the refraction of the tsunami of 27–28 March 1964: The Crescent City case. Proceedings, Ocean Science and Ocean Engineering 1965–Joint Conference Marine Technology Society and American Society Limnology and Oceanography, Washington, D.C. 2:707–716.

Rutherford, D. 1986. Disaster at Scotch Cap. *The Keeper's Log* 2:12–14.

Satake, K. 1988. Effects of bathymetry on tsunami propagation: Application of ray-tracing to tsunamis. *Pure and Applied Geophysics* 126:27–36.

Satake, K., and F. Imamura. 1995. Introduction to "Tsunamis: 1992–1994." *PAGEOPH* 145:373–379.

Satake, K., and H. Kanamori. 1991. Use of tsunami waveforms for earthquake source study. *Natural Hazards* 4:193–208.

Satake, K., K. Shimazaki, Y. Tsuji, and K. Ueda. 1996. Time and size of a giant earthquake in Cascadia, inferred from Japanese tsunami records of January 1700. *Nature* 379:246–249.

Schwing, F. B., J. G. Norton, and C. H. Pilskaln. 1990. Earthquake and bay: Response of Monterey Bay to the Loma Prieta earthquake. *Eos* 71:250–251, 262.

Self, S. 1992. Krakatau revisited: The course of events and interpretation of the 1883 eruption. *Geojournal* 28.2:109–121.

Shi, S., A. G. Dawson, and D. E. Smith. 1995. Geomorphological impact of the Flores tsunami of 12th December, 1992. In *Tsunami: Progress in prediction, disaster prevention and warning,* ed. Y. Tsuchiya and N. Shuto, pp. 187–195. Netherlands: Kluwer Academic Publishers.

Shimamoto, T., A. Tsutsumi, E. Kawamoto, M. Miyawaki, and H. Sato. 1995. Field survey report on tsunami disasters caused by the 1993 Southwest Hokkaido earthquake. *PAGEOPH* 145:665–691.

Shuto, N. 1995. Tsunami: Disasters and defence works in case of the 1993 Hokkaido-Oki earthquake tsunami. In *Tsunami: Progress in prediction, disaster prevention and warning*, ed. Y. Tsuchiya and N. Shuto, pp. 263–276. Netherlands: Kluwer Academic Publishers.

Shuto, N., and H . Matsutomi. 1995. Field survey of the 1993 Hokkaido Nansei-Oki earthquake tsunami. *PAGEOPH* 144:649–663.

Sievers C., H. A., C. G. Villegas C., and G. Barros. 1963. The seismic sea wave of 22 May 1960 along the Chilean coast (trans. P. Saint-Amand). *Bulletin of the Seismological Society of America* 53:1125–1190.

Singh, R. 1991. Tsunamis in Fiji and their effects. Prepared for the Workshop on Coastal Processes in the South Pacific Island Nations, Lae, Papua New Guinea, 1–8 October 1987. *SOPAC Technical Bulletin* 7:107–120.

Smit, J., A. Montarari, N. Swinburne, W. Alvarez, A. R. Hildebrand, S. V. Margolis, P. Claeys, W. Lowrie, and F. Asaro. 1992. Tektite-bearing, deep-water clastic unit at the Cretaceous-Tertiary boundary in Northeastern Mexico. *Geology* 20:99.

Sokolowski, T. J. 1991. Improvements in the Tsunami Warning Center in Alaska. *Earthquake Spectra* 7:461–481.

Soloviev, S. L., and Ch. N. Go. 1984a. *A catalogue of tsunamis on the western shore of Pacific Ocean.* Translation by Canada Institute of Scientific and Technical Information, National Research Council, Ottawa, Ontario, Canada. Originally published Moscow: Academy of Sciences of the U.S.S.R., Nauka Publishing House, 1974.

———. 1984b. *A catalogue of tsunamis on the eastern shore of Pacific Ocean.* Translation by Canada Institute of Scientific and Technical Information, National Research Council, Ottawa, Ontario, Canada. Originally published Moscow: Academy of Sciences of the USSR, Nauka Publishing House, 1975.

Spence, W., S. A. Sipkin, and G. L. Choy. 1989. Measuring the size of an earthquake. *Earthquakes and Volcanoes* 21:58–63.

Synolakis, C., F. Imamura, Y. Tsuji, H. Matsutomi, S. Tinti, B. Cook, Y. P. Chandra, and M. Usman. 1995. Damage, conditions of East Java tsunami of 1994 analyzed. *Eos* 76:257–264.

Talandier, J., and E. A. Okal. 1979. Human perception of T waves: The June 22, 1977, Tonga earthquake felt on Tahiti. *Bulletin of the Seismological Society of America* 69:1475–1486.

———. 1989. An algorithm for automated tsunami warning in French Polynesia based on mantle magnitudes. *Bulletin of the Seismological Society of America* 79:1177–1193.

Talandier, J., D. Reymond, and E. A. Okal. 1987. M_m: Use of a variable-period mantle magnitude for the rapid one-station estimation of teleseismic moments. *Geophysical Research Letters* 14:840–843.

Taniguchi, T., and D. T. Woo. 1961. The seismic wave casualties in Hilo, Hawaii. *Archives of Environment Health* 2:434–439.

Tanioka, Y., L. Ruff, and K. Satake. 1995. The great Kurile earthquake of October 4, 1994, tore the slab. *Geophysical Research Letters* 22:1661–1664.

Thomson, D. J., L. J. Lanzerotti, C. G. Maclennan, and L. V. Medford. 1995. Ocean cable measurements of the tsunami signal from the 1992 Cape Mendocino earthquake. *PAGEOPH* 144:427–440.

Thornton, I. 1996. *Krakatau: The destruction and reassembly of an island ecosystem.* 346 pp. Cambridge, Mass.: Harvard University Press.

Thorsen, G. W. 1988. Overview of earthquake-induced water waves in Washington and Oregon. *Washington Geology Newsletter* 16:9–18.

Tsuji, Y., T. Yanuma, I. Murata, and C. Fujiwara. 1991. Tsunami ascending in rivers as an undular bore. *Natural Hazards* 4:257–266.

Uchiike, H., and K. Hosono. 1995. Japan Tsunami Warning System: Present status and future plan. In *Tsunami: Progress in prediction, disaster prevention and warning*, ed. Y. Tsuchiya and N. Shuto, pp. 305–322. Netherlands: Kluwer Academic Publishers.

Vitosek, M. J. 1963. The tsunami of 22 May 1960 in French Polynesia. *Bulletin of the Seismological Society of America* 53:1229–1236.

Walker, D. A. 1995. Optical measures of tsunami "shadow": A new technique for assessing tsunamigenic potential? SOEST Contribution 4075, Hawaii Institute of Geophysics and Planetology Contribution 880. Honolulu: University of Hawai'i.

———. 1996. Observations of tsunami "shadows": Technique for assessing tsunami wave heights? *Science of Tsunami Hazards* 14:3–12.

Walker, D. A., and E. N. Bernard. 1993. Comparison of T-phase spectra and tsunami amplitudes for tsunamigenic and other earthquakes. PMEL contribution 1355, SOEST contribution 3160, and JIMAR contribution 93–261.

Walker, D. A., C. S. McCreery, and Y. Hiyoshi. 1992. *T*-Phase spectra, seismic moments and tsunamigenesis. *Bulletin of the Seismological Society of America* 82:1275–1305.

Ward, S. 1982. Earthquake mechanisms and tsunami generation: The Kurile Islands event of 13 October 1963. *Bulletin of the Seismological Society of America* 72:759–777.

Washington Highways. 1964. Tidal wave rips coast highways. *Washington Highways* 11:2–3.

Weischet, W. 1963. Further observations of geologic and geomorphic changes resulting from the catastrophic earthquake of May 1960 in Chile. *Bulletin of the Seismological Society of America* 53:1237–1257.

Weller, J. M. 1972. Human response to tsunami warnings, the great Alaska earthquake of 1964. In *Human ecology*, pp. 222–227. Washington, D.C.: National Academy of Sciences.

Whitmore, P. M. 1993. Expected tsunami amplitudes and currents along the

North American coast for Cascadia subduction zone earthquakes. *Natural Hazards* 8:59–73.

Whitmore, P. M., and T. J. Sokolowski. 1996. Predicting tsunami amplitudes along the north American coast from tsunamis generated in the northwest Pacific Ocean during tsunami warnings. *Science of Tsunami Hazards* 14: 147–166.

Whitney, H. M. 1868. On the earthquake and eruptions of 1868. In Recent eruptions of Mauna Loa and Kilauea, J. Dana, ed. *American Journal of Science*, series II, 46:112–115.

Wilkes, C. 1850. *Narrative of the U.S. Exploring Expedition During the Years 1838, 1839, 1840, 1841, 1842*. Philadelphia: Lea & Blanchard.

Wilson, R. M. 1927. Tide observations. *Monthly Bulletin of the Hawaiian Volcano Observatory* 15:44–45.

Wood, F. J., ed. 1966. *The Prince William Sound, Alaska, earthquake of 1964 and aftershocks.* Publication 10-3, Coast and Geodetic Survey. Washington, D.C.: U.S. Government Printing Office.

Wood, H. O. 1914. On the earthquake of 1868 in Hawaii. *Bulletin of the Seismological Society of America* 4:169–203.

Woods, M. T., and E. A. Okal. 1987. Effect of variable bathymetry on the amplitude of teleseismic tsunamis: A ray-tracing experiment. *Geophysical Research Letters* 14:765–768.

Yamashita, F. 1996. *Meiji Sanriku Tsunami* [commemorative brochure for the one hundredth anniversary of the Meiji Sanriku tsunami], 47 pp. Mirikucho, Japan: Meiji Sanriku Tsunami Committee.

Yeh, H. H. 1991. Tsunami bore runup. *Natural Hazards* 4:209–220.

Yeh, H., F. Imamura, C. Synolakis, Y. Tsuji, P. Lui, and S. Shi. 1993. The Flores Island tsunamis. *Eos* 74:369, 371–373.

Yeh, H., P. Liu, M. Briggs, and C. Synolakis. 1994. Propagation and amplification of tsunamis at coastal boundaries *Nature* 372:353–355.

Yeh, H., V. Titov, V. Gusiakov, E. Pelinovsky, V. Khramushin, and V. Kaistrenko. 1995. The 1994 Shikotan earthquake tsunamis. *PAGEOPH* 144:855–874.

Young, R. W., and E. A. Bryant. 1992. Catastrophic wave erosion on the southeastern coast of Australia: Impact of the Lana tsunamis ca. 105 ka? *Geology* 20:199–202.

Zayakin, Y. A., and A. A. Luchinina. 1990. *Catalog of tsunamis in Kamchatka,* 81 pp. Boulder, Colo.: National Geophysical Data Center.

Zetler, B. D. 1988. Some tsunami memories. *Science of Tsunami Hazards* 6: 57–61.

Suggested Readings

Curtis, G. D. 1992. Tsunamis—seismic sea waves. In *Natural and technological disasters: Causes, effects, and preventive measures,* ed. S. K. Majumdar, G. S. Forbers, E. W. Miller, and R. F Schmalz. Easton, Pa.: The Pennsylvania Academy of Science.

Dengler, L., and K. Moley. 1995. *Living on shaky ground*, 23 pp. California Office of Emergency Services, Federal Emergency Management Agency, and Humboldt State University.

Francis, P., and S. Self. 1983. The eruption of Krakatau. *Scientific American* 249:172–187.

Fryer, G. 1995. The most dangerous wave. *The Sciences* (July/August):38–43.

Griffin, W., J. Ambrose, L. Griffin, J. Kriezl, and L. Daniel, eds. 1984. *Crescent City's dark disaster.* 188 pp. Crescent City, Calif.: Crescent City Publishing Company, Inc.

International Tsunami Information Center. 1995. *Tsunami—the great waves.* N.p.

Korgen, B. J. 1995. Seiches. *American Scientist* 83:330–341.

Lander, J. F. 1993. *Tsunamis affecting Alaska 1737–1996.* 195 pp. NOAA Pub. KGRD 31. Washington, D.C.: Department of Commerce, National Oceanic and Atmospheric Administration.

Lander, J. F., and P. A. Lockridge. 1988. *United States tsunamis (including United States possessions) 1690–1988.* 265 pp. NOAA Pub. 41-2. Washington, D.C.: Department of Commerce, National Oceanic and Atmospheric Administration.

Lander, J. F., P. A. Lockridge, and M. J. Kozuch. 1993. *Tsunamis affecting the West Coast of the United States 1806–1992.* 242 pp. NOAA Pub. KGRD 29. Washington, D.C.: Department of Commerce, National Oceanic and Atmospheric Administration.

Lynch, D. K. 1978. Tidal bores. *Scientific American* 238:146–156.

McCredie, S. 1994. When nightmare waves appear out of nowhere to smash the land. *Smithsonian* 24:28–39.

Napier, A. Kam. 1997. Landslide! *Honolulu Magazine* (February):28–34.

Okal, E. A. 1994. Tsunami warning: Beating the waves to death and destruction. *Endeavour* 18:38.

Reid, S. B., 1997. Tsunamis. In *Introduction to Hazards*, 3rd edition. United

Nations Disaster Program–Department of Humanitarian Affairs, InterWorks.

Satake, K. 1992 Tsunamis. In *Encyclopedia of Earth System Science 4*. Academic Press.

Walker, D. A. 1993. Pacific-wide tsunamis reported in Hawaii from 1819 through 1990: Runups, magnitudes, moments, and implications for warning systems, 117 pp. School of Ocean and Earth Sciece and Technology Report 93-2. Honolulu: University of Hawai'i.

———. 1994. Tsunami facts, 93 pp. School of Ocean and Earth Science and Technology Report 94-03. Honolulu: University of Hawai'i.

Illustration Credits

American Geophysical Union: Figure 7.8

Anonymous, provided by the National Geophysical Data Center: Figures 1.2, 1.10, 2.5, 5.2, 9.4

Ian Birnie: Figure 3.2

Michele Bullock, courtesy of the Pacific Marine Environmental Laboratory, NOAA: Figure 9.7

E. Coteau, courtesy of the University of Hawai'i at Mānoa Library, Pacific Collection: Figure 2.12

George Curtis: Figure 9.14

Amy Cutler: Figures 2.4, 2.7, 2.11

Lori Denger, Expedition with Brian Atwater: Figure 9.5

Walter Dudley: Figures 2.14, 3.4, 3.6, 8.1, 8.10, 9.11

Joe Halbig, provided by the National Geophysical Data Center: Figure 7.6

Harper's Weekly, provided by the National Geophysical Data Center: Figure 2.6

Henry Helbush, provided by the National Geophysical Data Center: Figure 4.3

Hokusai: Figure 2.1

Honolulu Star-Bulletin, courtesy of the International Tsunami Information Center: Figure 4.5

Honolulu Star-Bulletin, courtesy of the Joint Institute for Marine and Atmospheric Research: Figure 4.7

Joint Institute for Marine and Atmospheric Research, University of Hawai'i at Mānoa, courtesy of George Curtis: Figures 1.20, 2.13, 4.8

Herb Kawainui Kāne: Figure 7.2

Karakuwa Tsunami Taikenkan, courtesy of Fumio Yamashita: Figure 9.3

Look Laboratory, University of Hawai'i: Figure 9.9

Ted Lusby, provided by the National Geophysical Data Center: Figure 1.11

Gordon A. Macdonald, Agatin T. Abbott, and Frank L. Peterson, *Volcanoes in the Sea: The Geology of Hawaii* (second edition), pp. 318–319, copyright 1983 by University of Hawai'i Press: Figures 2.16, 5.17

Charles L. Mader: Figure 9.10

Alexander Malahoff, courtesy of the Hawaii Undersea Research Laboratory: Figure 7.10

Maruzen Co., Ltd., courtesy of the Joint Institute for Marine and Atmospheric Research: Figure 5.12

Don J. Miller, courtesy of the U. S. Geological Survey: Figures 2.8, 2.9, 2.10

J. G. Moore and G. W. Moore: Figure 7.7

R. Nakamura, *Honolulu Star-Bulletin,* courtesy of the International Tsunami Information Center: Figure 4.6

National Academy of Sciences, from "Effects of Tsunamis: An Engineering Study" by B. W. Wilson and A. Torum, pp. 361–523. In *The Great Alaska Earthquake of 1964: Oceanography and Coastal Engineering,* 1972: Figure 6.3

National Oceanic and Atmospheric Administration (NOAA): Figure 7.12

NOAA/EDIS, provided by the National Geophysical Data Center: Figures 4.2, 6.1, 6.4, 6.5

Pacific Marine Environmental Laboratory, NOAA, courtesy of Eddie Bernard: Figures 9.6, 9.8

Pacific Tide Party, provided by the National Geophysical Data Center: Figures 5.15, 5.16

Pacific Tsunami Museum

- The *Hawaii Tribune-Herald* Collection: Figures 5.7, 5.8, 5.9, 5.11
- The Jeanne Branch Johnston Collection, Betty Mason Dease: Figure 1.12
- The Martha McNicoll Collection: Figure 1.13
- The Bunji Fujimoto Collection: Figure 1.5
- The Cecilio Licos Collection: Figure 1.9
- The Larry Nakagawa Collection: Figures 1.15, 1.16a
- The Shigeru Ushijima Collection: Figure 2.15

Pacific Tsunami Warning Center: Figure 3.3

Wayne Rasmussen, provided by the National Geophysical Data Center: Figures 1.6, 1.7

Pierre St. Amand: Figure 5.1

Dennis Segrist, provided by the National Geophysical Data Center: Figures 8.7, 8.8, 8.9

Seismological Society of America: Figures 5.3, 5.4, 5.5, 5.6

John Smith, SeaBeam Image courtesy of Alexander Malahoff, Hawaii Undersea Research Laboratory: Figure 7.11

Grace Walsh Tessier, courtesy of Shaun Fleming: Figure 3.1

Dorothy H. Thompson: Figure 7.4

Gordon Tribble, U. S. Geological Survey: Figure 7.9

Bruce Turner (photo of computer image by Paul Wessel), West Coast/Alaska Tsunami Warning Center: Figure 3.5

Takaaki Uda, Public Works Research Institute, Japan; taken by S. Sato; provided by the National Geophysical Data Center: Figure 8.2

University of California–Berkeley, provided by the National Geophysical Data Center: Figures 1.8, 1.14, 1.18, 1.19, 2.17

U. S. Army Corps of Engineers, provided by the National Geophysical Data Center: Figures 1.16b, 1.17, 5.13, 6.6

U. S. Coast Guard, provided by the National Geophysical Data Center: Figures 1.1, 1.3

U. S. Geological Survey, provided by the National Geophysical Data Center: Figure 6.2

U. S. National Park Service, provided by the National Geophysical Data Center: Figures 7.3, 7.5

U. S. Naval Historical Center, collection of Rear Admiral Luther G. Billings, courtesy of his daughter, Mrs. A. B. Hendrickson: Figure 2.2

U. S. Naval Historical Center, Murray Greene Day, New Bedford, Mass.: Figure 2.3

U. S. Navy, provided by the National Geophysical Data Center: Figure 4.1

W. G. Van Dorn, provided by the National Geophysical Data Center: Figure 5.14

Daniel Walker, Tsunami Memorial Institute: Figure 9.12

Paul Wessel: Figure 3.7

Fumio Yamashita: Figures 9.1, 9.2, 9.13

A. Yamauchi, *Honolulu Star-Bulletin,* courtesy of the International Tsunami Information Center: Figure 4.4

Harry Yeh, provided by the National Geophysical Data Center: Figures 8.3, 8.4, 8.5, 8.6

Index